Global Communication

This volume interrogates what "global" means in the context of "communication," and who benefits from global communication practices and industries. Emerging scholars contribute their unique perspectives in communication scholarship, charting innovative directions for research that connects empirical evidence with pressing questions of social significance.

Contributing authors question who benefits from global communication, moving beyond unbridled enthusiasm for the technological potential of digital media toward more serious and substantive empirical assessments of the underlying utility of these communications as processes. This critical reflection leads to considering problems that result from the way in which global communication becomes mobilized, in the practice of journalism and development, as well as the information and communications technology (ICT) industry.

Karin G. Wilkins serves as Professor of Media Studies, Associate Director with the Center for Middle Eastern Studies, and Chair, Global Studies Bridging Disciplines Program at the University of Texas at Austin, previously as Director of Media Studies. Wilkins has won numerous awards for her research, service, and teaching. Her work addresses scholarship in the fields of development communication, global communication, and political engagement.

Joseph D. Straubhaar is the Amon G. Carter Centennial Professor of Communications in the Department of Radio-TV-Film at the University of Texas at Austin. He has been Director of Media Studies for the RTF Department and the Director of the Center for Brazilian Studies within the Lozano Long Institute for Latin American Studies. His primary teaching, research, and writing interests are in global media and cultural theory, media and migration, digital media and the digital divide in the USA and other countries, and global television production and flow.

Shanti Kumar is an Associate Professor in the Department of Radio-TV-Film at the University of Texas at Austin. He is the author of *Gandhi Meets Primetime: Globalization and Nationalism in Indian Television* (University of Illinois Press, 2006) and the co-editor of *Planet TV: A Global Television Reader* (New York University Press, 2003). He has also published book chapters in several edited anthologies and articles in journals. His research and teaching interests include global media studies, cultural studies, Indian cinema and television, and postcolonial theory and criticism.

New Agendas in Communication

A Series from Routledge and the College of Communication at the University of Texas at Austin

Roderick Hart and Stephen Reese, Series Editors

This series brings together groups of emerging scholars to tackle important interdisciplinary themes that demand new scholarly attention and reach broadly across the communication field's existing courses. Each volume stakes out a key area, presents original findings, and considers the long-range implications of its "new agenda."

Agenda Setting in a 2.0 World
edited by Thomas J. Johnson

Identity and Communication
edited by Dominic L. Lasorsa and America Rodriguez

Political Emotions
edited by Janet Staiger, Ann Cvetkovich, and Ann Reynolds

Media Literacy
edited by Kathleen Tyner

Communicating Science
edited by LeeAnn Kahlor and Patricia Stout

Journalism and Citizenship
edited by Zizi Papacharissi

The Interplay of Truth and Deception
edited by Matthew S. McGlone and Mark L. Knapp

Global Communication

New Agendas in Communication

Edited by
Karin G. Wilkins, Joseph
D. Straubhaar, and Shanti Kumar

Routledge
Taylor & Francis Group
NEW YORK AND LONDON

First published 2014
by Routledge
605 Third Avenue, New York, NY 10017

Simultaneously published in the UK
by Routledge
4 Park Square, Milton Park, Abingdon, Oxon OX14 4RN

Routledge is an imprint of the Taylor & Francis Group, an informa business

Library of Congress Cataloging in Publication Data
Global communication : new agendas in communication / edited by Karin Wilkins, Joe Straubhaar, Shanti Kumar.
pages cm
1. Communication, International. 2. Mass media and globalization.
3. Mass media- -Social aspects. I. Wilkins, Karin Gwinn, 1962- II. Straubhaar, Joseph D. III. Kumar, Shanti.
P96.I5G535 2013
302.2–dc23
2013003602

ISBN: 978-0-415-82895-6 (Hardback)
ISBN: 978-0-415-82897-0 (Paperback)
ISBN: 978-0-203-74460-4 (ebook)

Typeset in Sabon
by Taylor & Francis Books

Contents

List of Contributors

Lisa Brooten (PhD) is an Associate Professor in the College of Mass Communication and Media Arts at Southern Illinois University Carbondale. Her research focuses on media democratization, local and global social movement media, community media, indigenous media, human rights, gender and militarization, and interpretive/critical research methods. Her recent research includes a comparison of media reform efforts in Thailand, the Philippines and Burma (Myanmar).

Paula Chakravartty is Associate Professor of Individualized Study, The Eallatin School and the Department of Media Culture, and Communication, New York University. Her teaching and research cover news media as well as telecommunication and information and communication technologies, science and technology studies, postcolonial and feminist political theory, and political economy, with a focus on India, Brazil, and the USA.

Florencia Enghel is a PhD candidate in Media and Communication at Karlstad University in Sweden. Her dissertation focuses on the relationship between communication, citizenship, and social change. She has been a guest lecturer and thesis supervisor at Malmö University's Master's program in Communication for Development. She has published "Teens Changing the World: Youth, Communication and Social Change" (in 2010, co-edited with Thomas Tufte, RUC, Denmark) and the special issue of *Nordicom Review*, "Mobilizing Global Communication: For What and For Whom?" (in 2012, co-edited with Karin Wilkins, University of Texas at Austin). She serves as Vice-Chair of the Participatory Communication Research Section of the International Association for Media and Communication Research (IAMCR).

Marwan M. Kraidy is Professor of Global Communication at the Annenberg School for Communication at the University of Pennsylvania. He has published six books, including *Reality Television and Arab Politics: Contention in Public Life* (Cambridge, 2010), which won the 2010 Best Book Award in Global Communication and Social Change from the International Communication Association, the 2011 Diamond Anniversary

Best Book Award from the National Communication Association, and the 2011 Roderick P. Hart Outstanding Book Award from the Political Communication Division, National Communication Association.

Shanti Kumar is an Associate Professor in the Department of Radio-TV-Film at the University of Texas at Austin. He is the author of *Gandhi Meets Primetime: Globalization and Nationalism in Indian Television* (University of Illinois Press, 2006) and the co-editor of *Planet TV: A Global Television Reader* (New York University Press, 2003). He has also published book chapters in several edited anthologies and articles in journals. His research and teaching interests include global media studies, cultural studies, Indian cinema and television, and postcolonial theory and criticism.

José María García de Madariaga is a Professor at the Universidad Rey Juan Carlos of Madrid (URJC), where he teaches subjects related to journalism, online democracy, and digitization. In 2003 he received his PhD in Sciences of the Information from the Universidad Complutense of Madrid with a dissertation about the counter-informative influences in mainstream media in the digital era. Since then, he has developed individual and collective research centered on the influence of digitization on the processes of social communication, especially in journalism and citizenship. Among them is his book *El periodismo en el Siglo XXI: Una profesión en crisis ante la digitalización* (*Journalism in the 21st Century: A Profession in Crisis Facing Digitization*) (Dykinson, 2008).

Aswin Punathambekar is Assistant Professor of Communication Studies at the University of Michigan, Ann Arbor. He is the author of *From Bombay to Bollywood: The Making of a Global Media Industry* (NYU Press, 2013) and co-editor of Global Bollywood (NYU Press, 2008).

Yeidy M. Rivero is an Associate Professor in the Department of Screen Arts and Culture and the Department of American Culture at the University of Michigan, Ann Arbor. Her work has been published in journals such as *Media, Culture & Society*, *Feminist Media Studies*, *Television & New Media*, *Global Media and Communication*, *Cinema Journal* and *Critical Studies in Media Communication*. She is the author of *Tuning out Blackness: Race and Nation in the History of Puerto Rican Television* (2005). Her current book project focuses on Cuban commercial television during 1950 to 1960.

Joseph D. Straubhaar is the Amon G. Carter Centennial Professor of Communications in the Department of Radio-TV-Film at the University of Texas at Austin. He has been Director of Media Studies for the RTF Department and the Director of the Center for Brazilian Studies within the Lozano Long Institute for Latin American Studies. His primary teaching, research, and writing interests are in global media and cultural theory, media and migration, digital media and the digital divide in the USA and other countries, and global television production and flow.

Karin G. Wilkins serves as Professor of Media Studies, Associate Director with the Center for Middle Eastern Studies, and Chair, Global Studies Bridging Disciplines Program at the University of Texas at Austin. Wilkins has won numerous awards for her research, service, and teaching. Her work addresses scholarship in the fields of development communication, global communication, and political engagement.

Introduction

New Agendas in Global Communication

Karin G. Wilkins, Joseph D. Straubhaar, and Shanti Kumar

Emerging trends in global communication shift how we connect socially, organize politically, and distribute resources economically. In order to explore new agendas for research, this volume interrogates what we mean by "global" in the context of "media" and "communication," and who benefits from global communication practices and industries. Emerging scholars offer "new agendas" in communication scholarship, charting innovative directions for research that connects empirical evidence with pressing questions of social significance.

First, we problematize what is meant by the term global in this context. With the increasing use of the term "globalization," it is even more imperative that we clearly define it through understanding the cultural geography of global, regional, national, and local media. Often it seems like everything is global, but we hope to interrogate that term closely to see which elements really are global in scope and impact, and which may be better thought of as regional, oriented around transnational cultural-linguistic spaces, specific to a certain kind of genre, or other analytical approaches that will be explored in this volume.

Critical evaluations of media production, distribution, and consumption practices within cultural contexts offer some insight into how we "mediate" the global. Chapters drawing attention to communications in Latin America, the Arab World, and South Asia contribute to this work by complicating territorial boundaries, exploring how local audience as well as industry practices work within global as well as local configurations. Moreover, the political and economic consequences of global communication need to be recognized. We question who benefits from global communication, moving beyond unbridled enthusiasm for the technological potential of digital media, toward more serious and substantive empirical assessments of the underlying utility of these communications as processes. This critical reflection leads us to consider problems resulting from the way in which global communication becomes mobilized in the practice of journalism and development as well as the information communication technologies (ICT) industry.

What Is and Isn't Global in Global Media?

One of the most common starting points in discussion of globalization in general is Appadurai's five landscapes of globalization (Appadurai 1996). Since one of the theoretical linchpins of this volume is cultural geography, landscapes of the global are a particularly good place to start. As Appadurai maps them, the five overlapping but disjunctive landscapes of globalization are:

1 ethnoscapes produced by flows of people: tourists, immigrants, refugees, exiles and guest workers; ...
2 technoscapes, the machinery and plant flows produced by multinational and national corporations and government agencies; ...
3 finanscapes, produced by the rapid flows of money in the currency markets and stock exchanges; ...
4 mediascapes, the repertoire of images of information, the flows which are produced and distributed by newspapers, magazines and film; ...
5 ideoscapes, linked to flows of images which are associated with state or counter-state movement ideologies which are comprised of elements of freedom, welfare, rights, etc.

Featherstone (1990, pp6–7)

For Appadurai, the suffix "scapes" reveals how the five related but sometimes disjunctive flows of finance, technology, migration, media, and ideology are deeply perspectival in the global cultural economy. In other words, global landscapes look different in different contexts. However, most theorists agree that financial flows have become truly global in reach and impact, moving economic resources (and crises) quickly from one place to another (O'Brien 1992), and that the substantial changes wrought by the virtually worldwide expansion of market capitalism (Wallerstein 1979) have created a changed economic infrastructure for most nations and places. This factor is particularly relevant to us, since most media globalization theorists would probably also agree that the overall globalization of economic systems has changed media powerfully, particularly in pushing media toward more commercial, private forms in most, if not all, countries and regions (Herman 2003; Straubhaar 2007).

Furthermore, most students of media globalization would probably agree that the tremendous changes in global technology infrastructure have had genuinely global impact, including videotape on television, which permitted television programs to flow between countries (Nordenstreng and Varis 1974); videocassette recorders, which permitted people to move television and film across borders with fewer controls and more individual choice (Boyd and Straubhaar 1985); direct satellite TV and audio broadcasting, which let thousands of new channels stream over the borders of countries from India to Turkey, forcing existing channels to compete with foreign or transnational ones (de Sola Pool 1979); and most recently, the Internet, which permits increased media flows (Cunningham 2001), new forms of transnational activism (Dagron 2001), and new forms of interaction among people, non-governmental organizations (NGOs), businesses, governments, etc. (Castells 2001). So it is not hard to conclude that there has been real global change in the technologies underlying the media, access to them, and creation of content for them. Similarly, the scholars who look at the rise of diasporic media (media created for transnational migrants or media from "back home" used by them) would agree that the acceleration of migration between parts of the world as part of globalization has transformed the need for and role of media for those migrating (Georgiou 2006).

Somewhat ironically, however, we do intend in this volume to complicate quite a bit how the globalization of mediascapes occurs in what Appadurai (1996) called the global cultural economy. Taking into account the foregoing arguments about the impacts of global economics, technology, and migration, it is hard to argue that media are embedded in a changed global system. Even here, however, there are many second thoughts about what the various facets of the term global actually mean. Cultural, linguistic, and political geographies other than the global are often very powerful, despite these economic and technological changes. Although China is arguably moving from being a system of socialism with special characteristics to one of capitalism with special characteristics, it is hard to conclude that Chinese and American capitalism are becoming similar, much less homogenous, despite their tight global linkages through trade. Similarly, early technological optimists thought that the decentered, networked nature of the Internet would trump all national efforts to control transnational communication through it (Kalathil and Boas 2003); but China succeeds in doing just that to a substantial degree. While assumptions about the power of technology *per se* in globalization are problematic, it is equally problematic to dismiss what technology can enable. Despite government controls on overtly political material, the Internet in China does permit millions of young Chinese to download American television shows such as *Friends* when the government does not permit it to run on the airwaves. So the dialectic of technological possibility, economic change, government actions, and individual agency involved in global media is complex.

This, in fact, is our starting point, that global systemic changes have reshaped the possibilities for media in most places; but knowing what has actually developed out of the new possibilities is considerably more complicated. Rather than thinking that the global reach of technologies or media capitalism has necessarily determined much, we will instead try to see what new possibilities are opened up or enabled, and who has the effective access and resources to use those possibilities.

Global Communication Defined by Cultural Geography

Much of the best current work on global media and communication tends to define it in terms of what might be called cultural geography. In Chapter 1 of this volume, Straubhaar maps a cultural geography for media where some media are truly global in their reach and impact, a large and increasing number are transnational or regional but not truly global, many remain defined primarily by national boundaries, a smaller but increasing number address cultural or linguistic regions within nations, and some are defined by the cities in which they are centered, even though these are media capitals (Curtin 2007) with regional, transnational, or even global reach, such as Hong Kong or Beirut. Straubhaar defines several kinds of truly global production, flow, and reception. These include the global production and export reach of Hollywood, seen by some as a new kind of networked or media-based empire (Hardt and Negri 2001, 2004). New work on Hollywood tends to see it less as the direct reflection of US national interests, as it was earlier seen by Schiller (Schiller 1969), and more now as the center of a global network of finance, co-production, and distribution that pulls the film industries and film-makers of many cultures into working with Hollywood groups (Miller et al. 2005).

The truly global media also now include genre-oriented film and television groups, such as National Geographic, Animal Planet, and Discovery Channel (Chris 2002; Mjos 2010), which have evolved genre forms, like entertainment-oriented documentaries, that can easily be co-produced, dubbed, or adapted across cultural and linguistic borders. These tend to circulate through specialized, globally branded television channels on satellite or cable. Perhaps of even higher impact in their broader circulation and attendance are the wide variety of TV formats (Moran 1998), from reality shows such as *Big Brother* or *American Idol*, to scripts and production packages for melodramas such as *Ugly Betty*. The transnational and/or global export and licensing for these formats for producing television redefine TV flows away from a strict focus on the flow of programs or channels to the flow of ideas, genres, and formats for producing television, films, music, and other forms of communication locally or nationally but within global forms, which blurs one of the key classic definitions of global media as those which are exported across national or other cultural borders. The discussion of media production, flow, and reception that is transnational but not exactly global began in the 1980s with the observation of increased television flow within geographic regions, such as Latin America (Straubhaar 1991; Wilkinson 1995), where Wilkinson was the first to note that the key element of the new pattern was not so much regional/geographic, but cultural-linguistic. Similar trends were observed in Asia (Chan 1996; Sinclair et al. 1996) and the Middle East (Amin 2000; el-Nawawy and Iskander 2002). Straubhaar (2007) argues that there are two primary emerging cultural geographies of transnational television: geo-cultural, where countries share a common region, a history of cultural interaction with neighbors, and often shared colonial experiences, language, religion, and ethnicity—prime examples would be Latin America, East Asia, South Asia, and the Middle East; and cultural-linguistic spaces or markets, where common colonial history and language is shared across considerable distances, as in the Anglophone world (the UK, the US, Canada, Australia, etc.) and the Francophone and Lusophone worlds, based on the wide-ranging British, French, and Portuguese empires (see Chapter 1).

Other scholars are making a case for a translocal landscape of interactions in which globalization takes place within and across fairly specific localities. With the advent of satellite TV in the 1990s, for example, new competitive television situations were created in the Middle East (Kraidy 2005; see also Chapter 2 of this volume) and in India (Kumar 2006), among others, by television channel groups based in other areas: Hong Kong in the case of India, Western Europe in the case of the Middle East.

While Kraidy's chapter in this volume (Chapter 2) details the situation in the Middle East, the case of India provides an interesting parallel. When Rupert Murdoch's Hong Kong-based STAR TV network made rapid forays into the Indian television markets in the early 1990s, many feared the rise in transnational satellite television represented a new phase of Western media imperialism in the subcontinent. However, in order to compete with other foreign and domestic networks that were also trying to cater to the diversity of linguistic, cultural, and regional interests of audiences in India, STAR TV strategically embraced a "think global, program local" policy. As part of this "glocalization" strategy, STAR TV launched a music network called Channel [V] that targeted Indian youth audiences by programming in a hybrid language of Hinglish: at once, both Hindi and English (Kumar and Curtin 2002).

In 1993, Murdoch's NewsCorp also acquired a stake in a Hindi language network Zee TV to cater to large segments of Indian audiences—particularly in north India—who primarily watched television programs in Hindi on the state-owned network Doordarshan. By the mid-1990s, Zee TV was the most popular Hindi-language network on satellite television in India, as well as in other markets around the world, such as the UK, Africa, and North America. By September 1999, when Zee TV bought out Murdoch's stake in the network, it was clear that the local network Zee had emerged stronger from the deal than the global giant NewsCorp, which few would have initially anticipated. Following the success of Zee TV, other networks such as Sony TV and STAR TV also began programming in Hindi. Soon, other commercial networks (e.g., ETV and Sun TV) started programming in regional languages such as Telugu and Tamil. By the end of the 1990s, transnational channels such as STAR TV and regional language channels such as ETV, Sun TV, and Zee TV had all expanded into multilingual networks, programming in several languages ranging from Tamil, Telugu, Kannada, and Malayalam to Marathi, Punjabi, Gujarati, and Bengali, to name a few (Kumar 2006).

The growth of multilingual programming on transnational networks such as STAR TV and on translocal networks such as Eenadu TV, Sun TV, and Zee TV in India reveals that the landscape of global television is not a homogenous terrain, and that globalization cannot be equated with Westernization or

Americanization of the world. Instead, globalization of media in any specific context can only be understood in relation to diverse and disjunctive flows simultaneously emanating from and to the transnational, national, translocal, and regional levels.

The Global, the National, the Regional, and the City

All of these global, translocal, regional, and transnational landscapes or layers of media are clearly part of global communications, if that is broadly defined. In this volume, we also argue that nations, regions within nations, and global cities or media capitals are also crucial parts of what might be seen as a global or world system of media. Indeed, before the conversation began to focus on global communication, many scholars and practitioners spoke of international communication: communication between nations (Thussu 2009). The case has been strongly made that in many parts of the world, nations are still the most powerful actors (Morris and Waisbord 2001). Many powerful states can still define and regulate domestic economic markets, regulate media and advertisers, and reinforce national identity through textbooks, national holidays, sports, etc. Many global media and advertising corporations, in fact, find it useful to accept those defined national market boundaries and work within them (Straubhaar 2007).

However, all of these concepts and forces are in interaction and dialogue. Curtin (2005) has argued that although China is a powerful nation state, able to take on and essentially radically limit Rupert Murdoch's efforts to enter the Chinese market, it also must be understood as part of a transnational landscape of cultural production that includes Taiwan, Singapore, and the overseas Chinese (Curtin 2007), as well as including global media capitals such as Shanghai and Hong Kong that have their own direct connections with global markets (Curtin 2003).

In Chapter 3 of this volume, Yeidy Rivero discusses how Fidel Castro made sophisticated use of the commercial television networks developed in Cuba before 1959 by local and US businesses before he proceeded to consolidate power and a socialist system by nationalizing those same media. Punathambekar and Kumar (Chapters 4 and 5 of this volume, respectively) also point to the contradictions of globalization that exist within and beyond national boundaries of India, and in the Indian film industry known as Bollywood. In different ways, these chapters illustrate that the globalization of media is not simply a one-way flow from a pre-constituted space called the global located in Western capitals such as London or Paris, or American media capitals such as Hollywood or New York City, to "other" locations around the word (often marked as non-Western, peripheral, or Third/developing world). Instead, they illuminate how the "global" is constituted through the very practices of media production, circulation, and distribution in diverse locations around the world, even as those "local" practices are themselves transformed through the process of globalization.

This process has been described by Stuart Hall (1997) as the articulation of the global and the local in media, communication, and culture. Hall borrows the concept of articulating the global and the local from Laclau and Mouffe (1985, p105), who define articulation as the process of "establishing a relation among elements such that their identity is modified as a result of the articulatory practice." In terms of the globalization media and communication, articulation can help to explain why neither the "global" nor the "local" has any fixed meaning, or any predetermined relationship across time and space. Instead, the meanings and relationships of the global and the local can only be constituted through their articulation or linking to particular places, cultures, or contexts.

Subsequent chapters in this volume engage this critical reflection in considering who benefits and who gets hurt from the mobilization of global communication. In the context of development, large bilateral and multilateral institutions benefit through their work in terms of attracting resources to support their organizational agendas, though whether and how communities designated as "beneficiaries" actually benefit should be questioned. In Chapter 7, Enghel engages this critique through her exploration of United Nations (UN) discourse and praxis, expanding on Wilkins's Chapter 6 that argues for a shift in global communications away from development studies toward more comprehensive and critical work in "advocacy communication." Brooten's Chapter 8 on how social movements engage the media to work through "transversal" spaces offers an important direction along these lines, interrogating the work of communities engaged in collective resistance across and in spite of sovereign state boundaries.

How communications technologies traverse global spaces needs to be examined not only to document the potential success of social movements in mobilizing participants and communicating arguments, but also to remind us of potential political and economic limitations. In Chapter 9, Chakravartty recognizes the constraints imposed through corporate control of the information technology (IT) industry in India. This political-economic analysis points to a series of critical macro-level questions that we should engage in global communications work, whereas de Madariaga's Chapter 10 on how journalists use communications technologies in their work points to the organizational practices and professional interests that are guided and constrained through global processes.

Attention to new agendas in global communication begins with questioning the very premises of our field. What, indeed, does "global" mean in relation to "communication?" How does our very focus of our conceptualization of global communication structure our observations? And what does this imply for our research agendas? If we want to suggest that global communications means a capacity for people to mobilize across space and time in ways that are distinct from recent historical dynamics, then we need to consider the value as well as the problems encountered.

New research agendas in global communication should engage in more critical reflection based on empirical work. We want to know more about the organizational dictates that encourage some and discourage other uses of globalized communication within a context of political agencies, such as governments and advocacy groups, as well as economic giants, such as corporations.

References

Amin, H. Y. (2000) "The current situation of satellite broadcasting in the Middle East," http://www.tbsjournal.com/Archives/Fall00/Amin_Paris.htm, accessed 19 October 2001.

Appadurai, A. (1996) *Modernity at Large: Cultural Dimensions of Globalization*. Minneapolis, MN: University of Minnesota Press.

Boyd, D. A. and J. D. Straubhaar (1985) "The developmental impact of the home videocassette recorder in the third world," *Journal of Broadcasting and Electronic Media*, 29(1): 5–21.

Castells, M. (2001) *The Internet Galaxy: Reflections on the Internet, Business and Society*. Oxford, UK: Oxford University Press.

Chan, J. M. (1996) "Television development in greater China: Structure, exports, and market information," in J. Sinclair et al. (eds) *New Patterns in Global Television: Peripheral Vision*. Oxford: Oxford University Press, 1–25.

Chris, C. (2002) "All documentary, all the time? Discovery Communications Inc. and trends in cable television," *Television & New Media*, 3(1): 7–28.

Cunningham, S. (2001) "Popular media as public 'sphericules' for diasporic communities," *International Journal of Cultural Studies*, 4(2): 131–147.

Curtin, M. (2003) "Media capital: Towards the study of spatial flows," *International Journal of Cultural Studies*, 6(6): 202–228.

Curtin, M. (2005) "Murdoch's dilemma, or 'What's the price of TV in China?'," *Media Culture Society*, 27(2): 155–175.

Curtin, M. (2007) *Playing to the World's Biggest Audience: The Globalization of Chinese Film and TV*. Berkeley, CA: University of California Press.

Dagron, A. G. (2001) *Making Waves: Story of Participatory Communication for Social Change*. New York, NY: The Rockefeller Foundation.

de Sola Pool, I. (1979) "Direct broadcast satellites and the integrity of national cultures," in K. Nordenstreng and H. Schiller (eds) *National Sovereignty and International Communication*. Norwood, NJ: Ablex.

el-Nawawy, M. and A. Iskander (2002) *Al-Jazeera: How the Free Arab News Network Scooped the World and Changed the Middle East*. Cambridge, MA: Westview Press.

Featherstone, M. (1990) "Introduction," in M. Featherstone (ed) *Global Culture*. London: Sage Publications.

Georgiou, M. (2006) *Diaspora, Identity and the Media: Diasporic Transnationalism and Mediated Spatialities*. Cresskill, NJ: Hampton Press.

Hall, S. (1997) "The local and the global: Globalization and ethnicity," In A. D. King (ed) *Culture, Globalization and the World System*. Minneapolis, MN: University of Minnesota Press, 19–40.

Hardt, M. and A. Negri (2001) *Empire*. Cambridge, MA: Harvard University Press.

Hardt, M. and A. Negri (2005) *Multitude: War and Democracy in the Age of Empire*. London: Penguin Books.

Herman, E. S. a. R. M. (2003) "The rise of the global media," in L. Parks and S. Kumar (eds) *Planet TV: A Global Television Reader*. New York, NY: New York University Press.

Kalathil, S. and T. C. Boas (2003) *Open Networks, Closed Regimes: The Impact of the Internet on Authoritarian Rule*. Washington, DC: Carnegie Endowment for International Peace.

Kraidy, M. (2005) *Hybridity or the Cultural Logic of globalization*. Philadelphia, PA: Temple University Press.

Kumar, S. (2006) *Gandhi Meets Primetime*. Urbana, IL, University of Illinois Press.

Kumar, S. and M. Curtin (2002) "Made in India," *Television & New Media*, 3: 345–366.

Laclau, E. and C. Mouffe (1985) *Hegemony and Socialist Strategy*. London: Verso.

Miller, T., N. Govil et al. (2005) *Global Hollywood 2*. New York, NY: Pearson Longman.

Mjos, O. (2010) *Media Globalization and the Discovery Channel Networks*. New York, NY: Taylor & Francis.

Moran, A. (1998) *Copycat Television: Globalisation, Program Formats and Cultural Identity*. Luton, Bedfordshire, UK: University of Luton Press.

Morris, N. and S. Waisbord (eds) (2001) *Media and Globalization: Why the State Matters*. Lanham, MD: Rowman & Littlefield.

Nordenstreng, K. and T. Varis (1974) *Television Traffic—A One-Way Street*. Paris: UNESCO.

O'Brien, R. (1992) *Global Financial Integration: The End of Geography*. New York, NY: Council on Foreign Relations Press

Schiller, H. I. (1969) *Mass Communication and American Empire*. Boston, MA: Beacon Press.

Sinclair, J. S., E. Jacka et al. (1996) *New Patterns in Global Television*. New York, NY: Oxford University Press.

Straubhaar, J. D. (1991) "Class, genre and the regionalization of the television market in Latin America," *Journal of Communication*, 41(1): 53–69.

Straubhaar, J. (2007) *World Television: From Global to Local*. Thousand Oaks, CA: Sage Publications.

Thussu, D. K. (2009) "Internationalize media studies," *Television & New Media*, 10(1): 162–164.

Wallerstein, I. (1979) *The Capitalist World Economy*, Cambridge, MA: Cambridge University Press.

Wilkinson, K. (1995) *Where Culture, Language and Communication Converge: The Latin-American Cultural Linguistic Market*. Austin, TX: University of Texas.

Chapter 1

Mapping "Global" in Global Communication and Media Studies

Joseph D. Straubhaar

This chapter explores the changing cultural geography of globalization as a multiplicity of levels of media use and identity. It argues that even before the era of globalization, people have often acquired layered identities that were local, regional, transnational—based on geo-cultural or cultural-linguistic regions—and national (Anderson 1983). People around the world continue to strongly reflect these layers or aspects of identity, while many also acquire new layers of identity that are transnational or global.

These increasingly multilayered identities are articulated with a complex variety of changing structures (Hall 1997). As we shall see below, economies, political powers, social class, and geography structure who can access what new channels. Media institutions themselves are becoming more complexly multilayered as they reach further geographically. Institutional models, such as commercial TV networks, globalize, but are also localized and regionalized as they engage the specific histories and institutions of a variety of cultures, media traditions, and regulatory systems. Identities also layer up as people migrate, acculturate, live abroad, travel, learn languages, join or leave religions, and, although the experiences are less directly personal and less intense, perhaps as they acquire access to a variety of forms of media.

Cultural Sedimentation: Layers and Mixtures/Hybrids

In my own work, particularly in a number of in-depth interviews within Brazil, and in Texas, from Austin to the Texas border with Mexico, I find a process of the formation, maintenance, occasional collapse, and recent proliferation of multiple layers of identity and corresponding layers of media use. I frequently find myself interviewing people who articulate their identity as connected to a series of geographic spaces or cultural or social layers. In both Brazil and Texas, I found people who articulate very clearly a local sense of self, a state/province or regional sense of self, an ethnic and/or linguistic sense of self, a national sense of self, varied interest or knowledge of the global, and, most often, a gendered sense of self, social class sense of self, a religious sense of self, and a professional and educational sense of self.

So, in my work, I am moving toward what we might see as a kind of sedimentary model in which layers of meaning, culture, identity, and media use form and persist. New layers form over the top of all others as structural circumstances permit or even dictate. Sometimes when we look at a culture, for instance, we are likely to see the newest layer as strongest. When many observers look at cultures these days, they see on top a new layer of what they might call globalization. Some might suppose that this is perhaps now the dominant layer, perhaps homogenizing all the others—or perhaps even the dominant aspect of someone's identity or experience. However, if we look at a highway or a river canyon someplace, we see the layers from the side. We realize that there are recent layers, which are important, but that they've built up over older layers, which are still there.

Particularly in reference to culture, we think we will see that these layers interact with each other so that as one becomes globalized in a certain part of one's life, such as one's education or profession, one finds that change inevitably in dialogue with one's local interests, musical preferences, or even religious or philosophical interests. One could have a rich metaphor to work with for the ongoing nature of change. Too many systems either overstate change, as in the very strong cultural imperialism hypothesis, or, in contrast, understate change, as in some globalization hypotheses in which everything forms separate pockets and resists or simply persists. I don't think any of these are quite adequate for understanding the complexity of what we see: People do use new forms of media to interact, forming new layers and levels of identity, but they also continue to think in terms of older layers of identity, interest, and media use.

Articulation of New Spaces and Layers of Identity

Hall (1997, p44) observed that "the great collective social identities, which we thought of as large-scale, all-encompassing, homogenous, as unified collective identities," such as class, race, nation, gender, and "the West," are still present and efficacious, but are no longer complete master concepts. He saw such collective identities more as processes of identification, never completed, related to evolving perceptions of difference between groups. He cited the example of how many young people wanted to be Caribbean, Black, and British, not just one or the other (Hall 1997). Hall's recognition of the increasing multiplicity of identities and identifications is one of the bases of my argument about multiple levels and spaces of identity.

Hall also argued that identities did not grow or flow freely: that they were articulated to texts, to institutions, and to societal structures of economic, social, and political power. While identities change, multiply, and have relative autonomy, they are also tied to sources of power, to economic structures and class positions, to producers of media (Hall, 1980), but not in simple or reductionist ways. We would like to build on this idea to see how different levels of identity are articulated to changes in media access structured by class, to cultural capital that guides and limits choices and understandings among media offerings, and to other forces that can be seen as articulated themselves through complex cultural geographies.

The Importance of Cultural Geography

Building on Michel de Certeau, I argue that a cultural geography, like a city, is constructed both from above and below. De Certeau notes that strategic decisions are made top down by political leaders, planners, and industrialists, creating spaces that vary from rural areas filled with sugar plantations originally worked by slaves (a political-economic decision that still defines much of the northeast coast of Brazil) to cities organized for commerce, finance, or industry. However, those who live in and use these spaces make tactical decisions on the ground that can, to some degree, redefine them by what the place's users do with the space, what practices they evolve (de Certeau 1984).

Multiple Identifications and Multiple Media Spaces of Production and Flow

People increasingly identify with multiple cultures at various layers or spaces. People identify with multiple cultural groups or symbols in different fields of activity. People establish different identities at school and work, in sports or religion, with family and friends. They form multiple layers of cultural capital specific to a field of activity (Bourdieu 1984). People form different dispositions to behave differently with various groups, corresponding to different levels or spaces of identity. Someone can be religious and traditional in Spanish at home with their parents, adventurous and critical in either English or Spanish with different friends, sports-minded in Spanish with some friends, and achievement-oriented in a specific transnational English vocabulary with colleagues at work. All of these different layers of identity and culture will have varied connections to global, cultural-linguistic, national, and local spaces and forces.

These are based in varying combinations of cultural geography, institutional strategies and alliances, and cultural productions based on genre linked to institutions, nations, and other cultural spaces. Complex cultural geography is based on cultural definitions of markets and spaces for production, flow, and consumption. Institutional power is defined by geography, but also redefines cultural geography, offering new cultural products with which to identify.

The following is an examination of a series of levels of identity or identification with cultural flows and spaces. Some of this comes from examining the production process, some from extensive mapping I recently conducted of television flows (Straubhaar 2007), and much comes from interviews with a variety of audience members in Salvador and São Paulo, Brazil, and central-south Texas that I conducted with colleagues and students at the University of Texas. Interviewing in São Paulo mostly took place during 1989 to 1998; in Salvador, during 1990 to 1998, and 2003 to 2006; and in Texas, during 1998 to the present. In Texas, students and I conducted interviews, primarily with Latinos, but also Anglos and African-Americans in East Austin, Texas, during 1999 to 2005 and in 2009, and in South Texas, during 2004 to 2008. During 2004 to 2005, and 2009, that interviewing focused on Mexican migrants to the US.[1] Both semi-structured interviews about media and less structured life and family histories with media were gathered. Rantanen (2005) provides a very engaging example of how carefully examined family histories of media use may inform analyses of globalization.

Cultural Geography: Cultural Distance; Global, National, and Local Identities

Let us begin to focus clearly here on the experience and understanding of the audience. All audiences have a strong sense of local identity, as they live their daily lives in a specific place and time. The dominant cultural forces for most people seem to have been originally local and regional. People thought of themselves primarily in terms of villages, local dialects, perhaps tribes or clans. Only after 1700 did most Frenchmen speak French, in one of the earliest nations to be defined as such (Weber 1976). That sense of locality lingers.

Based on my interviews in Brazil, even the most globe-hopping businessman or academic usually still has a very *local identity*, based on the specific cultural geography of a neighborhood they live in, or other neighborhoods they go to for work or pleasure. That locality may be defined by planners on maps but is often more vividly defined by restaurants, music clubs, nightclubs, museums, libraries, bookstores, and coffee shops: all the places that in a very physical and spatial sense tend to give life local context and local pleasure. These are linked to personal networks, but in media terms, also local music scenes, local radio, local newspapers, festivals, and performances. Their local interpretive communities (Lindlof et al. 1988) which mediate both national and local media (Martín-Barbero 1993) are based on families, local friends, neighbors, clubs, church congregations, sports groups, etc.

Most of their day-to-day thought and interaction is still very local, as will be seen in the Brazilian examples below. So, much of the knowledge or cultural capital that people use in interpreting media, such as television, comes from the very local experience of daily life in a very specific culture shaped by their immediate environment, place, and time. Sometimes the experience of locality sharpens a sense of difference between the experience of local life and what is shown of global or national scenes on television. For example, rural Brazilians observed and interviewed by La Pastina (2003) felt themselves to be on the periphery of Brazil, somewhat alienated from what they saw of urban Brazilian life on television.

The experience of locality is bound up with other layers of identity, such as ethnicity or language/culture. For example, Rosemary Alexander Isett found that Inuit interviewees in Alaska felt considerable distance from the culture of the lower 48 US states as portrayed on television. In fact, watching US culture on television, they tended to be relieved that they lived in a rural area within their own ethnic culture (Isett 1995). Latinos interviewed in East Austin also tended to think in terms of their identities in terms of very local conditions, particularly neighborhoods and extended families. Like the Inuit, they often felt considerable cultural distance from the dominant US, both in terms of language and culture. However, this sense of cultural distance varies greatly between generations, which will be discussed further below, and families. In fact, our fieldwork with Latinos has brought in the layer of family identity very powerfully. Family with Latinos often roots one in a specific neighborhood, but can also connect oneself to extended family back in Mexico as well. So while family is often bound up with locality, it is different with migrants, particularly those who see themselves as part of a temporary diaspora, with plans to permanently, or cyclically, return to what one considers a cultural homeland, such as Mexico or Central America.

However, cultural unity at the local level (if it ever existed) has been broken by distanciation (Giddens 1990). As that concept implies, local people begin to identify at a distance with non-local cultural trends and peoples through media, schooling, travel, work, migration, and the vicarious experience of others in their personal networks. However, this is less new than many theorists of globalization suppose. Almost all cultures have been in relatively constant contact with others and hybridization between cultures has been more the norm than not (Pieterse 2004). A more interesting question for us here might be how much various forms of media have accelerated the relative weight and pace of distanciation, change, and hybridization. We want to give weight to the relative solidity of the local, as many audiences experience it, without falling into what Pieterse (2004, p47) calls "the reification of the local, sidelining the interaction between the local and the global."

Well before the nation state achieved relative primacy in institutional definitions of identity, local cultures generally had cultural exchange and interpenetration with larger cultural linguistic regions. Local groups were part of larger language, culture, and ethnic groups, like the Inca, which had trade, ties, and affiliations with many smaller, distinct groups across much of what is now Peru, Bolivia, and Chile, or the Mayan cultures which sprawled across what is now southern Mexico and Guatemala. These pre-national cultures continue much of what is local or sub-national in many parts of the world. Even local cultures whose languages have been extinguished usually leave behind traces in food, agriculture, names, and ethnic physical characteristics that continue to structure cultural locality. This is the *longue durée*, or long-term process, of hybridity, as discussed in Pieterse (2004, p91). In this view, hybridity is an extremely old, long-run process, into which modern media fall.

Identity at the Level of State, Province, or Region

Many people have a sense of region that is larger than their immediate locale, but smaller than the nation. In larger nations such as Brazil and China, or even medium-sized nations such as Mozambique or Spain, there are usually coherent sub-national regional identities that reflect these enduring hybrid forms of identity. In many places, those conform to the cultural contours of pre-European/colonial cultures. As we shall see below, in the case of Brazil, these local and (sub-national) regional cultures tend to maintain a sense of difference from national cultures that mediates their use of national media (Martín-Barbero 1993). The experience of local identities in Brazil subsumes or even conflates ethnic differences, as different regions of the country have very different ethnic mixes. In other countries, such as Nigeria or India, regional differences might be at once linguistic, religious, and ethnic. People there tend to filter national, supra-national, and global media through this filter of perceived regional or local difference, even as they adapt elements of those outside cultures into the local aspect or level of their own multilayered identities.

For many people, there are *regional layers of identity* and regional layers of media, which are smaller than the nation, but larger than the very local. Many of these regional foci depend on language variation. Very important cultural and linguistic regions, from Catalans in Spain to the Kurds in Turkey, to Mayan language speakers in southern Mexico to a variety of regional language groups in India, still have strong layers of quite separate identities from the nation states that contain them. They also often have ethnic, cultural, and religious differences with national majority populations. They usually have their own music traditions and scenes (O'Connor 2002), and histories, which are often intense focuses of identification. Sometimes hundreds of millions in very large nations such as India speak local or regional languages linked to regional cultures, film industries, broadcasters, etc.

In other large nations, such as Brazil, even when regions are not divided by language or religion, there can still be noticeable regional cultures. Although all speak Portuguese and the large majority is Catholic, there are noted regional identities that support regional cultural industries, particularly in music, in the south (Leal 1989), north (McGowan and Pessanha 1998) and northeast (Vianna 1999). So, regionalism varies with its bases of language, religion, ethnicity, and musical tradition in different countries, but is very widespread; and many people, including those whom I interviewed in Salvador, Brazil, reveal regionalism as a distinct layer of identity in their media and cultural choices when interviewed.

Geo-Cultural Identities

Another strong set of pre-national and pre-global cultural forces are *geo-cultural*, often based on cultural-linguistic groups that precede the European colonization beginning in the late 1400s. They tend to have strong geographic regional proximity, hence the geo-, that turned in lasting close cultural ties. These are largely based on ethnic groups, languages, empires, and religions in places such as Greater China, the Arab World, and South Asia that predate the European colonization that began during the 1400s. Some other cultural forces and identities that remain very powerful date from before most nation states, from colonial empires, migration, languages, religions, and racial mixtures in Latin America, Franco- and Lusophone Africa, in the US, and other Anglophone nations, such as Australia or Canada (Abram 2004). Post-colonial areas such as Latin America now share centuries of historical, cultural-linguistic, ethnic, dynastic or political, and religious roots of geo-linguistic (Sinclair 1999) or geo-cultural layers of understanding and identification (Straubhaar 2005). They were reinforced by the media of the time. Books have flowed among language and religious groups for millennia. The Bible and the Koran helped to expand large areas of shared religious identification and, at least in the case of the Koran, considerable Arabic language hegemony, as well, in the Middle East and North Africa. Music, drama, poetry, and other "media," broadly defined, helped to reinforce such ties, even before printed books.

Cultural-Linguistic Transnational Layers of Identity

There can also be meaningful associations that are geographically widely spread but linked by language, culture and history, colonial experience, and now academic interaction that builds on common literatures, administrative traditions, literary, dramatic and poetic forms, etc. There are more geographically far-flung groupings, such as the Portuguese-speaking world (Portugal, Angola, Mozambique, and Brazil, along with much smaller places such as East Timor) or the English-speaking world. I distinguish them from the geo-cultural by calling them *cultural-linguistic transnational layers of identity or spaces.*

Former French colonies still import from France and produce for themselves in French, creating a geographically dispersed transnational culture of *francophonie*, which supports film, television, and music production in a number of places, from Quebec to Senegal. Similarly, former Portuguese colonies are linked in a Lusophone culture, or *Lusofonia*, which centers on both Brazil and Portugal, but mostly Brazil in terms of television production, although it also extends to East Timor and a number of African countries. Former British colonies are similarly linked, and like the Lusophone cultures, television production is dominated not by the former colonial power, but by a larger former colony, the US. For these audiences, as well as those who speak Arabic, Chinese, Hindi, Persian, Turkish, etc., their populations have migrated globally enough to create diasporic groups that constitute large, commercially attractive transnational audiences.

While the roots of these geo-cultural and transitional cultural linguistic layers of culture predated both the nation state and globalization, they were reinforced by many of the new forces, particularly in technology and economics, that we associate with globalization. Just as we see growth in Hollywood coverage of the world (Miller et al. 2005), we see rapid growth in cultural-linguistic and geo-cultural spaces and markets in television exports (Straubhaar 2007), satellite/cable TV (Sinclair et al. 1996), Internet sites, and music and movie downloads.

National Identities

In most places with relatively coherent, relatively powerful nation states, people adopt a layer of identity which reflects an imagined national identity, as proposed by Anderson (1983). Historically, that process was experienced as beginning with the incursion of national political institutions, such as the army, tax collectors, and governors, into local places and cultures; but in many places those institutions left a relatively light footprint on a still largely local sense of identity.

Urban elites have often been shocked at how little involved many rural and small town people are with the nation. Correspondingly, those same urban elites have often radically misunderstood the local communities and identities constructed by people in the nation's periphery. When Peruvian novelist Vargas Lhosa was looking for a metaphor or precedent to understand the enigmatic rural rebellion of the supposedly Maoist Shining Path in Peru (Vargas Llosa 1981), he looked at the striking example of a completely misunderstood religious community in rural Brazil in the 1890s whose withdrawal into a mountain stronghold in southern Bahia at Canudos was (mis)interpreted by the national leadership as a monarchist revolt inspired by France, which led them to send several successive armies to level the place and movement, creating one of the classic, enduring narratives of regional versus national identity in Brazil (da Cunha 1973).

As Gramsci, (1971, p33) and others note, the sense of nationality among the broad population tended to increase in most places with government-organized or supervised schooling. When schooling is left to religious groups, as it was in much of Latin America before the 1900s, or as it still is in large parts of the Middle East (extending to Afghanistan and Pakistan), then national government loses at least some control over a principal "ideological state apparatus" (Althusser 1992) for defining identity and setting ideology in its own desired directions. As Anderson points out, print capitalism, combined with the efforts of the state to further define national identity in the 19th century (at least in Europe and Latin America) through newspapers and novels, provided readers with a sense of common activity, purpose, and identity with those across the same political and linguistic boundaries (Anderson 1983). Anderson shows that national cultures and national media are not a given and are slowly and often carefully constructed over time. During the 20th century, it seems that radio and television have continued that process in many countries, extending it to less literate, more remote peoples within the national boundaries (Katz and Wedell 1976). Sometimes governments have deliberately employed cultural elements. During the 1930s, for example, the samba and other music in Brazil was used to articulate national identity (Vianna 1999).

Still, while national identities seemed to become the dominant or at least primary form of identity to many in the 20th century, we would argue that what was created was a new layer of identity above locality and, in some cases, above sub-national cultural-linguistic communities, supplementing but not necessarily reducing those older layers. However, national power over media still remains strong in most of the world, at least where powerful, coherent governments persist (Morris and Waisbord 2001). Many people during the last two centuries have interacted with modern national educational systems in which textbooks and teaching norms are discussed and prescribed at a national level. They have also grown up with national media. They frequently deal with other nationalizing forces, such as national churches, national labor unions, national sport teams, and institution-led patriotism, as well as all the things that in many ways are articulated with a certain sense of nationalism.

Language- and Culture-Defined Transnational Spaces and Markets

Viewers can feel both distance and proximity to what they see on television. Cultural geographic distance relates to viewers' perceived position on the periphery, as well as to cultural proximity to a national culture to which they are exposed at multiple levels. While their perceived cultural distance is very real, so is the lifelong cultivation of national identity that they receive through television, schooling, national music on the radio, national sports such as soccer, national symbols such as the flag, etc.

Cultural proximity is based on a number of factors that work to overcome the cultural distance described above. Cultural proximity is achieved at a number of levels: local, where shared experience is directly physical and experiential; regional, where shared experience can be either direct or mediated; national, where the community is largely imagined through the media (Anderson 1983); and transnational, where identities are based on shared language and history at a more removed, but perhaps still quite real, level.

Transnational cultural proximity is based largely on language. However, besides language, there are other levels of similarity or proximity based on cultural elements: dress, ethnic types, gestures, body language, definitions of humor, ideas about story pacing, music traditions, religious elements, etc. that are often shared across national borders. Indian movies are popular in the Arab world for such similarities; and Brazilian *telenovelas* (evening serials or soaps) dubbed into Spanish are more popular than the American *Dallas* or *Dynasty* because of such similarities (Straubhaar 1991). Iwabuchi (1997, 2002) shows that young Taiwanese people tend to see Japanese television and music as culturally proximate, sharing a sense of "Asian modernity" despite the language difference between Japanese and Chinese.

The creation of national borders and identities did not radically reduce existing supra-national cultural-linguistic communities and identities. Most people in China, Hong Kong, Singapore, and Taiwan all thought of themselves as Chinese, even though those nations and autonomous provinces, as defined in the 20th century, were often in violent ideological or even military conflict. The same is probably less true, interestingly enough, among Spanish speakers in Latin America, where early dreams of pan-regional identity by the original liberators, such as Bolivar, faded under waves of nationalism and conflict over borders throughout the 19th and 20th centuries. But those countries had not had four to five previous millennia of cultural development before the conflicts of independence, as was the case with Chinese culture. Four centuries of common Spanish language and heritage brought peoples together, which we still see in the cultural-linguistic space in which television moves around Latin America; but cultures were also differentiated from the beginning by varied patterns of *mestizajé* (miscegenation), hybridization, and cultural layering among many indigenous groups, Spanish and Portuguese colonizers, imported African slaves, and more recent European immigrants, such as the Italians and Germans who greatly changed the character of Argentina, Chile, southern Brazil, and Uruguay.

A strong linguistic base still unites cultural markets in Spanish-speaking Latin America with Spain, and Portuguese-speaking Brazil with (not only) Portugal, but also parts of Africa and Asia. A similar geographically dispersed pattern from the British Empire unites the US with Australia, Canada, the UK, Jamaica, and New Zealand. If that strikes US readers as unlikely, think about which few countries the US does import movies and music from.

New Global Layers of Cultural Identification

People around the world also acquire new layers of identity or identification corresponding to new global layers of production and flows of media, enabled by new structural forms of political economy, as well as new forms and models of media. Forming one of the main new global layers, Hollywood dominated the flow of film to most world markets (Miller 2001) and, at least initially, dominated the flow of television as well (Nordenstreng and Varis 1974). So by sheer dint of exposure, American culture began to seem as a familiar second culture to many people (Gitlin 2001), particularly in Western Europe and the Anglophone countries where the US presence was often most notable, as well as around much of the rest of the world (Straubhaar 2007). US exports have also been prominent over the years in satellite/cable TV channels, music, and the Internet. A recent turn in political economy research calls this a new sort of virtual empire by the US (Hardt and Negri 2001). However, many people around the world were more lightly touched by this US layer than others, depending on location, social class, language, religion, and other aspects of identity that led them to discount the US output and choose to watch, read, and listen to other things (Hoskins and Mirus 1988).

There are also renewed and expanded transnational layers based on the roots noted above of cultural-linguistic and geo-cultural levels of production, flow, and identity. These utilize the same technologies as does the US media and cultural industries, contributing to the counter to US hegemony potential in what Hardt and Negri (2005) call the "multitude" of those who use the same mechanisms to either oppose the US or simply develop their own power within their own spheres, like Islamist forces within the Islamic world. Mass media such as radio, cinema, and television helped to consolidate and renew language and cultural groupings. Since the mid-1990s, research by Sinclair and others (1996) has highlighted the growing importance of cultural linguistic markets. In several cases, new export powers grew up within cultural-linguistic regions originally dominated by others, such as the US within the Anglophone world (Abram 2004) or Brazil within the Lusophone world (Marques de Melo 1988; Cabecinhas 2006). Research on the role of satellite TV in the Arab world (Kraidy 2002) shows how new refinements in media technology can continue to facilitate and reinforce such identities.

In television, the US continues to export heavily; but there are also new layers of other global production and flow or access, such as the worldwide but limited exports of Latin American *telenovelas*, Japanese *anime* on film and television, Hong Kong kung fu film and television, and Bollywood films, among others. Brazil alone is estimated to have exported *telenovelas* to over 140 countries. The scope of these flows, especially compared to Hollywood exports, has been contested (Biltereyst and Meers 2000), but they have grown into visible new options for those who have access to them. Some, such as Japanese *anime*, have grown beyond the contested limits of what were originally called counter-flows (Thussu 2007).

There is also a new global flow of news (with new operations such as Al-Jazeera in English), feature films (especially those of Bollywood or those co-produced with Hollywood), music, and some Internet sites, such as YouTube, where non-US material is growing rapidly (Cohen and Küpçü 2007). There are new media forms that seemed to be global in ways that we have perhaps not yet thought how to articulate, such as the way in which young people in many countries now interact in English via online role-playing games such as *World of Warcraft* or virtual worlds such as *Second Life*. These are now becoming so globalized that there are calls in 2009 for research proposals from several US government agencies who are trying to understand them as a new global form of communication.

Globalization and Audience Choices

It is easy to identify global corporations that are trying to operate in large parts of the world. Groups such as those owned by Rupert Murdoch have ambitions to own satellite/cable TV channels everywhere. These corporations powerfully reframe the political economy of how television works everywhere via ownership, advertising, and programming ideas. For an actual audience member, sitting in Brazil, China, or Mozambique, however, it is not always as easy to identify with the programming of such global groups. Most people in most places tend to primarily watch locally or nationally produced programs, not those of Hollywood or global producers such as Murdoch.

However, while people may be watching programs that are locally produced, this does not mean that they don't have a notable global or regional connection. The mass audience in most places tends to see quite a bit of "glocal" programming, globally inspired or even licensed, but nationally or regionally produced. While global flows of actual programs are very visible, particularly via satellite and cable channels, they just don't attract a majority of the audience in places where local, national, or regional material is available. Sparks (1998, 2005) observed that although CNN seemed to many to be an emerging space for a potential global public sphere, the actual audience for CNN International is, in fact, so small that it cannot have a very large impact. Audience members in countries large or rich enough to produce quite a bit of programming are probably exposed more to locally produced versions of global genres and formats, such as *Big Brother*, than to direct flows of television from abroad. For example, Brazilians see very little foreign television programming in prime time, but they are on their fifth version of *Big Brother*, which has been very popular.

So audience members worldwide are now exposed to two distinct levels of global influence. One is the direct flow of television programs first highlighted by Nordenstreng and Varis in 1974, as well as the direct flow of complete channels via cable and satellite first highlighted by Schiller (1981), Mattelart and Schmucler (1985), and others. Another is the copying or licensing of foreign formats, which brings varied levels of cultural influence and ideology (such as a focus on aspiring to and taking on consumer roles) with them (Fuller 1992). That process of glocalization has for some time taken place informally, as when format ideas for soap operas flowed both globally and regionally (see Chapter 10). It has also now taken a more formally licensed and traded form with the increasingly global traffic in packaged formats, such as *Big Brother* or *How to Marry a Millionaire*.

Multiple Levels of Audience Identity and Cultural Choices

This chapter also focuses on the lived experience of audiences with various levels and forms of world television. In particular, it focuses on their identities and how they form multiple identities that correspond in many ways to the multilayered cultural geography of television in the world. This seems to be due, in part, to the fact that many of those levels of experience with cultural geography are broadly felt by almost everyone in the television audience. All individuals have experience of their city or locality, which is increasingly addressed directly by local television production in many large cities or provinces. Although some people live isolated from any experience of nationality, the large majority of people seem to be touched by national television as well as nationally defined schooling, newspapers, books, radio, government, political campaigns, etc. This is clear in the Brazilian interviews discussed below.

The other central argument of this chapter is that while one may refer to identity for many people as increasingly hybrid, it is perhaps more clearly seen as multileveled. In fact, the two ideas are complementary. Hybridity is a long-term process in which all identities are constantly changing (Pieterse 2004); but we argue here that most people experience identity with regard to media, in part, as a series of cultural geographic levels from local to global. They also experience it, depending on the place and group, as related to language, culture, ethnicity, and religion. In addition, identities and media uses are divided still further by ethnicity and gender. So the idea of levels of identity *per se* is a heuristic device for understanding the way in which people seem to experience identity, particularly in relation to the complexity of world television.

Based on our review of the literature, as well as fieldwork in Texas and Brazil, we find that people make sense of media first through a set of cultural identities based on space and place: local, regional, national, and global. Related to this are identities based on culture and language, which tend to be linked to space and place, but can be transnational (e.g., transnational, post-colonial audiences, such as Francophones, Anglophones, or Lusophones, or the cases of migrants or diasporic communities). These identities can define television viewing. They can be maintained and recreated even after people have moved away from the place where the cultural linguistic identity was first formed (Appadurai 1996).

The next level seems to be one of social class based on the idea of cultural capital, linked to experiences within the family, school, neighborhood, and social groups, shaped by economic class commonalities (Bourdieu 1984) and a sense of group experience that solidifies into a class habitus (Bourdieu 1984). Whether they are actively aware of it or not, most people also experience and interpret media through a sense of ethnicity or race, as it is culturally constructed in the place and time where they live.

Almost all people similarly experience media through an awareness of themselves and their identity in terms of gender. Many people have a distinct sense of themselves in terms of race or ethnicity. In some countries, people also tend to have a very distinct sense of identity related to their age group. In the US, for example, media use and understanding are heavily segmented by age group, unlike in other societies. In Brazil, upper-class youth are likely to have media uses that are very different from their parents' uses; but working-class youth are less likely to show such significant differences (as reflected in interviews by the author). Finally, almost everyone experiences media refracted through a sense of values shared with a community, which is most often expressed in terms of religious group identification, but relates to other communities of values as well.

Depending on the circumstances of the individual and the groups with whom they are associated, different levels of identity might be of primary importance for experiencing and making sense of television and other media. For some, class identity might seem primary, as in some of the studies of the British working class by Morley and others (Morley 1992). For others, as in studies related to the reaction to some Hollywood productions by audiences in the Islamic Middle East, religion might be a primary identity for interpreting media experiences (Chambers 2002). For yet others, a primary identity for interpreting media might be ethnicity or gender, or both. For example, US Latinas tend to view imported Mexican television in terms of their own experiences of class, gender, and ethnicity within the Latino minority in the US (Rojas 2001).

Multiple Identifications

In this emerging model, people tend increasingly to identify with multiple cultures at various layers and levels. People can identify with multiple cultural groups in different fields of activity. There are different identifications that people establish at school and work, with family and friends. In this process of learning from others, people form multiple layers of cultural capital, often specific not only to a field of activity as Bourdieu (1984) would predict, but to different sub-cultures or cultural layers. People form different dispositions to behave differently with various groups. Someone can be religious and traditional at home with their parents, adventurous and critical with friends, sports-minded with team-mates, and achievement oriented with colleagues at work.

So within individuals, multiple cultural capitals coexist, specific to different fields of interaction and activity. Cultural elements survive or even thrive as layers of material practices among social classes, or as identities within individuals and collectivities in these classes. Although Bourdieu (1984) tended to see class or group habitus as singular, it seems likely that with these increasingly multiple cultural identities and practices, people may well have multiple habitus for different fields and situations.

All of these different layers of identity, culture, and even class habitus will have varied connections to global, cultural-linguistic, national, and local spaces and forces. While few people will reflect all these layers, I found in interviews that many people were, indeed, quite multilayered in both identity and media use. For example, one of the people whom I interviewed in Salvador, Bahia, in 2004 and again in 2005 had a number of layers of identity corresponding to levels of attention to media and information. He was a former hotel accountant, now working as a taxi driver because it paid better. He paid some attention to global issues via Brazilian television news and newspapers; he knew enough about the war in Iraq to want to make jokes about President Bush, but did not pay that much attention to a broad range of global issues. He was familiar with quite a bit of US popular culture, mostly in music and feature films; the only US television program he could remember was *The Simpsons*. He was somewhat conversant with a few bits of the Lusophone world; he knew about the recent independence of East Timor, he knew some fragments of Portuguese history, but he mostly made jokes about the Portuguese when asked about it. (Many Brazilians tend to tell jokes about the Portuguese very similar to jokes often told in the US about Polish immigrants.) He knew relatively more about Latin America in terms of news and events, knew some songs from other Latin American countries, and sometimes watched Mexican soap operas with his wife, who liked some of them (broadcast on SBT, the second most popular Brazilian network). He spent a lot of time with Brazilian national media, particularly newscasts, talk shows, *telenovelas*, sports, and music. He was passionate about national and local politics, talking about current national corruption scandals and giving local examples. He also knew a lot about the music of northeastern Brazil, in general, and the music of Bahia, specifically. He considered himself an evangelical Christian, but also seemed to know a lot about the Afro-Brazilian religion *candomblé* and some of the local music groups that played music from it. He considered himself lower middle class, but sympathized a lot with the poor and working class, and talked in terms of class issues quite a bit. He had some college-level training in accountancy, and clearly enjoyed talking with people whom he considered to be educated, but got along very well, and talked a lot, with fellow drivers with much less education, too. He is clearly of Afro-Brazilian descent and is proud of that in cultural terms, but wary of talking about it in anything related to politics. He liked to talk about Afro-Brazilian music, even in those terms, but was unwilling to talk about other aspects of racial issues, such as quotas for racial affirmative action, which are just being implemented in some Brazilian universities.

Overall, his education probably permits him a greater level of complexity in terms of identity and media use than most Brazilians, but he is hardly a member of the global elite; he is consciously someone struggling to stay in the lower reaches of the middle class. So he is perhaps somewhat typical of what a large number of Brazilians would aspire to be and to do in terms of identity and media use. (See Chapter 3 about the Brazilian preoccupation with defining and achieving middle-class status.)

Multilevel Identities and Social Class

One of the forces that drives and complicates multiple identifications at the local, national, cultural-linguistic, and global levels is stratification by social class within societies. The elite are the most globalized, as we will see below in our Brazilian cases. They have more access to multiple forms of media and information and communication technologies (ICTs). They accumulate a more globalized set of cultural capital through travel, education, work, and personal networks. For middle and working classes, their cultural capital is mostly national or regional/cultural-linguistic in terms of their media access, schooling, travel, work, etc. The cultural capital and life experience of the working class, poor, and homeless is mostly local. (See Chapter 3 for both theoretical examples and examples based on fieldwork in Brazil.)

Nevertheless, all of these class formations are also complex cultural groups with multiple identifications: local, national, cultural-linguistic, and global. Cultural elements survive or even thrive as layers of material practices among social classes and other groups, and as identities within individuals and collectivities in these classes. In the UK, Morley (1980) found people with working-class identities and cultural practices that created a sense of distance or critical viewing when they watched television, particularly BBC programs created from an elite point of view. Based on class identities, Morley found evidence of the active viewing/"reading" of television predicted by Hall (1980). Working-class people did critically accept, negotiate, or sometimes reject the intended readings that UK producers had put into their programs.

Nevertheless, a single interpretation of viewing and identity, based on class, cannot explain the complexity of what I have found in my interviews. People identify with multiple cultural groups in various fields of struggle (Bourdieu 1984). The idea of field is useful since people engage with multiple fields of activity and interest. Even within television, for example, people engage with the field of entertainment quite differently than the field of news (Benson and Neveu 2005). In Brazil, for example, we found people inclined to be more critical of television news, since they were aware of its censorship and manipulation, than of *telenovelas* even when they were produced by the same broadcaster, TV Globo, the dominant network in Brazil.

However, digging deeper in my last two years of interviews in Salvador, a major city in the more Afro-Brazilian part of the Northeast (during 2004 to 2005), I found that a number of people were turning away from TV Globo's *telenovelas*, saying that they could not relate to them as well as they could to other kinds of programs on other networks. For example, I found a recurrent phrase, "I don't see people like me on Globo," in interviews in Salvador. That phrase came up in several separate interviews. In an interview in 2004, a working-class Afro-Brazilian taxi driver used that phrase to explain why he was increasingly watching SBT instead of TV Globo. SBT, the second-rated national network, has had an explicit strategy of targeting lower middle-class, working-class, and working-poor viewers across Brazil ever since the 1980s, when its management began to realize that they could not compete across the board for the general audience (Fadul 1993). SBT's advertisements in media professional magazines clearly stress an audience defined by class, but do not make explicit that this working-class and poor audience will be largely Afro-Brazilian, given the strong demographic correlation that exists between class and race in Brazil (Telles 2004).

I asked the man if he meant that Globo did not have enough Black people on screen and that SBT had more. He said that maybe that was part of it, but wasn't able to elaborate much more; in fact, he seemed uncomfortable talking explicitly about race, as were several others when I asked them what they meant by comments similar to "I don't see people like me on Globo." They had a much easier time talking about how the people on TV Globo were always too rich, not like the people they knew. And they were able to articulate a sense of how Rio, where most of TV Globo's *telenovelas* and other programming is set, was a very different place than Salvador.

I have gone back over what people seemed to mean by this phrase. It seems to entail a varying combination of three layers of identity that emerged in the interviews. First, people are actively and consciously aware that most of the people portrayed on television, particularly on TV Globo, are much richer than they are. My interviewees openly articulated a sense of class difference with what they saw on television. Second, they are openly aware of the same distance, based on cultural geography, which La Pastina (2003) found in more rural parts of the northeast of Brazil, that those people on screen live in a very different part of the country with a substantially different culture (see above). Third, which only a few articulated openly, those people on screen seemed to be much "whiter" than viewers in Salvador, where most people are Afro-Brazilian.

Layers of Identities as Mediators

In this chapter we have reviewed a number of the primary layers of identity that influence television viewing, particularly among people we interviewed in Brazil and among Latinos in Austin. People's senses of identity in terms of cultural geography, class, ethnicity, and gender guide their choices of what to watch on television. Those viewing choices are the most basic and perhaps most important ways in which people interact with television. Multiple layers of identity also guide the ways in which people make sense of television, as well as how television seems to affect them.

Layers of identity can be mediating forces that help people to make sense of what they watch on television. In La Pastina's (2003) work in a rural Brazilian community, he found first that people expressed a sense of cultural distance from the television productions from the south of Brazil on TV Globo. That sense of their placement of a cultural geography of significant difference between the local and the national corresponds to what I found in interviewing in Salvador, Bahia, as well. Likewise, gender and people's differing levels of cultural capital were central mediating factors influencing various individuals' readings, interpretations, and occasional appropriations of content of some *telenovelas* that have been studied by La Pastina and others.

For example, racial identity, while muted among many Brazilian viewers, can provide a clear frame of interpretation, or mediator, for some Brazilians in their reactions to television content. In 2005, I watched a TV Globo song contest reality show called *FAMA* with a group of people at the Salvador, Bahia, school building about an Afrocentric music group for women called

Banda Feminina Didá. Their lead singer was competing for one of three spots for the mortheast of Brazil with seven other people. Three of ten semi-finalists were Afro-descendent, although the Black or mixed percentage of the region is well over 60 percent. One Black man whom the Didá people called a *negão* (handsome Black man) and two White people won. As the other regions of Brazil (north central and south southeast) also voted, the founder of the school, a well-known musician nicknamed Neguinho do Samba, became disgusted with the clear predominance of White singers in the winning circle. He looked at me and said: "Look, Professor, at the bias that is still there in this country." He was mad enough about it to walk out of the room, visibly irritated.

These senses of identity, which enable critical viewing, are also a form of cultural capital, another of Bourdieu's (1984) key ideas. We usually think of cultural capital as a set of socially approved knowledge and dispositions that relate to social mobility, based on things people learn from their families, schools, churches, and other major social institutions. For example, Bourdieu's original research found a strong connection between cultural capital acquired at home or school and taste for erudite versus popular music.

However, there are multiple layers of cultural capital, and individuals' related dispositions to behave in certain ways. In Bourdieu's (1984) terms, these kinds of formations of capital and dispositions are shared by groups in patterns that he refers to as the habitus of the group. In this case, we argue that individuals participate and identify with multiple class/group habitus, depending on which layer of identity and media experience we are talking about. For example, while watching television with several of the same people from the Banda Feminina Didá a year earlier in 2004, the television program denounced inadequate medical care for poor people. Most of the people shown were Afro-descendent, but not all. The commentary in the room at that point focused on the viewers' sense of solidarity with other poor or working-class people, such as the ones on screen in hospitals and clinics. One could infer racial solidarity, too, since most of the people shown were Afro-descendent; but it was not the main issue at the moment, even though it was clearly the primary issue for them a year later when the same people observed racial bias in the song contest. So, people clearly move among multiple senses of identity as frames of reference, as resources for understanding media, and as mediators of media content.

The idea of mediation has become a key theoretical principle in Latin American work about television, based on the research of Martín-Barbero (1993), who argued that we should move our focus from the media themselves to the forces that mediate our experiences with and understandings of media content. Both Martín-Barbero (1993) and Mexican media ethnographer Orozco Gómez (1996) argue for the importance of considering multiple sites of mediation. I would argue that this means consideration of mediation by multiple layers and sources of identity.

Much of this mediation process is very local and interpersonal. So as we have seen above, in cases such as La Pastina's study of Macambira, local culture can be a significant mediator of national, regional, or global television viewing. However, this local process is also informed by previous exposure to television, education, travel, and language learning, to advertising, other media such as books and newspapers, and a variety of personal experience, which creates cultural capital that mediates television viewing. Much of that cultural exposure, notably television programming, advertising, and education, is most often framed by a national process, guided in many nations by explicit state policy, so a dialectic between local and national levels of identification and cultural capital acts as a mediator of each other and of the global.

Multiple Identifications, Identity, and Hybridity

There is an ongoing, complex interaction between forces of economics and technology, as exemplified by many of the satellite television services, broadcast television networks, and Internet companies or institutions discussed above, and long-run patterns of culture and language. To a significant degree, people in audiences come to identify with what they are shown. The extraordinary dominance of US global film distribution since the 1920s has resulted in cultural patterns of familiarity, knowledge, and liking for American-style films that persists in many parts of the world (Miller 2001). This creates a market defined by both political economy and culture that new networks of feature and documentary film, such as HBO or Discovery, can exploit, using new technologies of television distribution. These supply and reinforce the audiences of those who like such genres. So, for large numbers of people, a specific identification with Hollywood-style films builds up to where there is a layer of culture so familiar to people that Gitlin called this American cultural layer of production, flow, and consumption a familiar second culture for many people in the world (Gitlin 2001).

To take a very different example, the historical primacy of public service broadcasting over time in the Nordic countries and its ongoing creation of genres and forms of content that engage and please its audience has created patterns of goodwill, familiarity, cultural capital or knowledge, and liking that continues to guide audience preferences toward it even when competition is available. From the continued ratings success of such cultural and informational genres in the face of both broadcast and satellite/cable multichannel TV competition, many Scandinavians seem to experience an ongoing identification with them that could be seen as fairly stable cultural layers of production, programming, and consumption for both informational and cultural forms typically identified with public service broadcasting. To some degree, these forms are identified with national culture and also continue to connect with and reinforce a layer of what might be seen as national identity in a country such as Denmark (Søndergaard 2003).

The creation of a certain linguistic or cultural space or market is intertwined with economic and technological forces. In his study of the development of nationalism, Anderson saw print capitalism as work with existing languages or dialects in order to standardize and spread them, via the printed word of newspapers, novels, etc., to become standardized national languages (Anderson 1983). In his work on modern India, Kumar (2006) shows how Hindi has both spread and been resisted as a national language within India by different institutions of television, at national and regional levels. By providing ongoing news and culture for people to identify with, a number of these broadcasters, at the regional or provincial level within India, have served to reinforce regional senses of identity, which were already based on earlier forms of language and culture, before television, radio, or film.

So, the interaction is, indeed, complex. Audience identification and more aggregated senses of cultural identity change with media forms. Culture is not static. Audience senses of identification can increase as forms of media bring them new and compelling cultural forms with which to identify. This is one of the ways in which layers of cultural production, flow, and identification can increase, reaching the multiple layers presented earlier in this chapter.

However, ongoing, changing forms of culture (and language) also define spaces and markets within which the use of technologies and orientations of media institutions and businesses is outlined. For example, there was a point in the history of Italian television broadcasting when, somewhat surprisingly, large audiences existed for Latin American *telenovelas*. As channels increased, seeking new material to program, programmers experimented with the *telenovelas* and struck a resonance or identification with parts of the audience. European scholars began to debate whether counter-flow from the developing to developed nations might be underway (Biltereyst and Meers 2000). However, an underlying preference for locally produced versions of popular television forms could also be seen or anticipated (Straubhaar 1991) and Italian fiction production began to increase, proving profitable, pushing the *telenovelas* slowly out of the main parts of the national programming schedule (Buonanno 2004). Still, for some parts of the Italian audience, particularly in southern Italy, where many felt linked with emigration and family ties to countries such as Argentina and Brazil, an identification with and liking for such programs continues (Del Negro 2003). These identifications with specific programs again reflect the growth of multiple layers of both identification and identity. These are not essentialized or reified, but must be seen in a steadily changing media and cultural environment where technologies, television institutions, program forms, and audience identification and identity evolve together.

This ongoing pattern of change can be seen as both hybridity and the multiplication of layers of production, programming/flow, and identification. The hybridity occurs in ongoing cultural change through the contact of local, regional, national, transnational, and global elements, such as those discussed earlier. Layers of cultural production and identification multiply as technological and economic forces allow. To people whom I have interviewed in Texas and Brazil, many of these layers of culture that are made available to them and with which they come to identify seem very solid, not something they anticipate changing. Latino immigrants to Texas whom I have interviewed, as well as Turkish immigrants to Western Europe interviewed in research by Ogan and Milikowski (1998), show that many immigrants welcome a certain continuity of culture to be found in television from back home. They cherish that layer of culture and identification, even as they form others in their new environment.

However, these layers of cultural production and flow evolve with technological and economic possibilities. Affordable satellite television channels make it much easier for transnational immigrants to stay more closely involved and identified with their home culture. (Earlier waves of immigrants had fewer media options and were more likely to have to use media in their new host's countries.) They also evolve with changing, or hybridizing, forms of culture that both reflect and frame the technological and economic possibilities. So, as television becomes cheaper, and people also start creating their own cultural forums on websites, we see the growth of Persian language television production in Los Angeles for Iranian immigrants there (Naficy 1993). We see even larger numbers of websites, web radio programs, and even specialized satellite TV channels for South Asian immigrants to the US or the UK, some focused on events back in South Asia, many focused directly on the immigrant experience and news of their own specific community (Mallapragada 2006). These examples show the reciprocity of economic, technology, culture, and media channels. People move in large numbers mostly for economic reasons, although political, familial, religious, and other reasons factor in as well (Papastergiadis 2000). As they move, they take their culturally formed interests with them. This creates spaces or markets for new layers of media to act in, if economic and technological possibilities allow. All of these ultimately tend to create a new layer of production, experience, and reception—that is, media, identification, and identity specific to the new immigrant community and its culture. Community and culture will thus represent both a hybridization of home and host cultures, and a new layer of media and culture in itself.

So, to conclude, this chapter looks at four bodies of issues and theory. First, we see the elaboration and development of new multiple layers of media production, flow, identification, and, perhaps eventually, identity. Second, these form in reciprocal interaction between technological possibilities, political-economic forces such as the movement of peoples and the expansion of media institutions and companies, and spaces for media created by the cultural identities and interests of concrete groups of people. The net effect of this has been the expansion of layers of production and reception of culture at the global, US export empire, transnational cultural-linguistic, transnational geo-cultural, translocal, national, regional, global city or media capital, and metropolitan and local levels.

Third, instead of the homogenization feared by earlier theorists (Hamelink 1983), we see a less drastic but perhaps equally pervasive hybridization of cultures. Both media institution professionals and audience members whom

I have interviewed tend to articulate what they see as the increase in the number of layers or kinds of culture (often expressed as new markets by the professionals). These layers are also constantly changing as they interact and hybridize over time. As a result, fourth, this is also a complex and dynamic system that is constantly evolving or emerging (Straubhaar 2007) as culture, political-economy, and technological possibilities interact and shape each other.

Note

1 The 2005 study was conducted under the supervision of Joseph Straubhaar, Viviana Rojas, Jeremiah Spence, Martha Fuentes, and Juan Pinon. It looked at family trajectories (occupation, education, migration) (González 2001; Maass and González 2005), personal histories (in-depth interviews and genograms), family histories, and assessment of media uses by generations, using surveys and semi-structured interviews. Interviews were conducted by 43 graduate and undergraduate students in two seminars. The sample was a convenience sample: families living in poor, working-class, and middle-class neighborhoods of Austin, Texas (for at least three years). There were a total of 34 informants: Latinos (24) and Anglos (10). We collected 18 individual life histories and another 16 from members of five multi-generation families (three Latino and two Anglo families). There were 22 females and 12 males. In terms of age, there were eight over the age of 50 (grandparents), fourteen from the ages of 30 to 50 (parents), and twelve from the ages of 18 to 29 (children).

The 2009 study conducted under the supervision of Joseph Straubhaar, Viviana Rojas, and Jeremiah Spence took a similar approach. Interviews were conducted by 58 graduate and undergraduate students in three seminars. We collected life histories from 23 two- to three-generation families (13 Latino, 2 Brazilian, 1 Korean, 5 Black, and 2 Anglo families).

References

Abram, D. J. (2004) *Four Tiers of Anglophone Television Importing: A Comparative Analysis of Television Flows between the United States, the United Kingdom, Australia, Anglo-Canada, Jamaica, Ireland and New Zealand from 1962–2001. Radio-TV-Film*. Austin, TX: University of Texas at Austin.

Althusser, L. (1992) *Ideologie et appareils ideologiques d'Etat* [Portuguese title: *Aparelhos ideologicos de Estado: nota sobre os aparelhos ideologicos de Estado*]. Rio de Janeiro, Brazil: Graal.

Anderson, B. (1983) *Imagined Communities: Reflections on the Origin and Spread of Nationalism*. New York, NY: Verso.

Appadurai, A. (1996) *Modernity at Large: Cultural Dimensions of Globalization*. Minneapolis, MN: University of Minnesota Press.

Benson, R. and E. Neveu (eds) (2005) *Bourdieu and the Journalistic Field*. New York, NY: Polity Press.

Biltereyst, D. and P. Meers (2000) "The international telenovela debate and the contra-flow argument," *Media, Culture and Society*, 22: 393–413.

Bourdieu, P. (1984) *Distinction: A Social Critique of the Judgement of Taste*. Cambridge, MA: Harvard University Press.

Buonanno, M. (2004) "Alem da proximidade cultural: Não contra a identidade mas a favor de alteridade" ["Beyond cultural proximity: Not against identity but in favor of alterity"], in M. I. Vassallo de Lopes (ed) *Telenovela—Internacionalização e interculturalidade*. São Paulo: Edições Loyola, 331–360.

Cabecinhas, R. (2006) "Identidade e Memória Social: Estudos comparativos em Portugal e em Timor-Leste," in M. L. Martins, H. Sousa, and R. Cabecinhas (eds) *Comunicação e Lusofonia: Para uma abordagem crítica da cultura e dos media*. Portugal: Campo das Letras, 183–214.

Chambers, D. (2002) "Will Hollywood go to war?" *Transborder Broadcasting Studies*, 8 (Spring/Summer), http://www.tbsjournal.com/Archives/Spring02/chambers.html.

Cohen, M. A. and M. F. Küpçü (2007) "Congress and the 'YouTube War'," *World Policy Journal*, 23(4): 49–54.

da Cunha, E. (1973) *Os sertoes*. São Paulo: Cultrix.

de Certeau, M. (1984) *The Practices of Everyday Life*. Berkeley, CA: University of California Press.

Del Negro, G. P. (2003) "Gender, class and suffering in the Argentinean *telenovela milagros*: An Italian perspective," *Global Media Journal*, 2(2).

Fadul, A. (1993) *The Radio and Television Environment in Brazil*. São Paulo: University of São Paulo.

Fuller, C. (1992) "Dutch reformat, resell shows," *Variety Europe*, 11 May: 75.

Gitlin, T. (2001) *Media Unlimited: How the Torrent of Images and Sounds Overwhelms Our Lives*. New York, NY: New York Metropolitan Books.

González, J. (2001) "Cultural fronts: Toward a dialogical understanding of contemporary cultures," in J. Lull (ed) *Culture in the Communication Age*. New York, NY: Routledge, 106–131.

Hall, S. (1980) "Encoding/decoding," in S. Hall, D. Hobson, A. Lowe, and P. Willis (eds) *Culture, Media and Language*. London: Hutchinson.

Hall, S. (1997) "Old and new identities, old and new ethnicities," in A. D. King (ed) *Culture, Globalization, and the World-System: Contemporary Conditions for the Representation of Identity*. Minneapolis, MN: University of Minnesota Press, 41–68.

Hamelink, C. J. (1983) "Cultural autonomy threatened," in C. J. Hamelink (ed) *Cultural Autonomy in Global Communications*. London: Longman Group, 1–25.

Hardt, M. and A. Negri (2001) *Empire*. Cambridge, MA: Harvard University Press.

Hardt, M. and A. Negri (2005) *Multitude: War and Democracy in the Age of Empire*. London: Penguin Books.

Hoskins, C. and R. Mirus (1988) "Reasons for the US dominance of the international trade in television programmes," *Media, Culture and Society*, 10: 499–515.

Isett, R. A. (1995) "Publicly funded satellite television in Alaska: Lost in space," *College of Communication Arts and Sciences*, Michigan State University: 274.

Iwabuchi, K. (1997) *The Sweet Scent of Asian Modernity: The Japanese Presence in the Asian Audiovisual Market*. Fifth International Symposium on Film, Television and Video, Media Globalization in the Asia-Pacific Region, Taipei.

Iwabuchi, K. (2002) *Recentering Globalization: Popular Culture and Japanese Transnationalism*. Durham, NC: Duke University Press.

Katz, E. and G. Wedell (1976) *Broadcasting in the Third World*. Cambridge, MA: Harvard University Press.

Kraidy, M. M. (2002) "Arab satellite television between regionalization and globalization," *Global Media Journal*, 1(1): 1–15.

Kumar, S. (2006) *Gandhi Meets Primetime*. Urbana, IL: University of Illinois Press.

La Pastina, A. C. (2003) *Viewing Brazil: Local Audiences and the Interpretation of the Nation*, Media in Transition 3. Cambridge, MA: MIT Press.

Leal, O. F. (1989) *Gauchos: Male Culture and Identity in the Pampas*. Berkeley, CA: University of California Press.

Lindlof, T. R., M. J. Shatzer et al. (1988) "Accommodation of video and television in the American family," in J. Lull (ed) *World Families Watch Television*. Newbury Park, CA: Sage Publications, 264.

Maass, M. and J. A. González (2005) "Technology, global flows and local memories Media generations in global Mexico," *Global Media and Communication*, 1(1): 105–122.

Mallapragada, M. (2006) "Home, homeland, homepage: Belonging and the Indian-American Web," *New Media and Society*, 8(2): 207–227.

Marques de Melo, J. (1988) *As telenovelas da Globo: Produção e exportação*. São Paulo: Summus.

Martín-Barbero, J. (1993) *Communication, Culture and Hegemony: From the Media to the Mediations*. Newbury Park, CA: Sage Publications.

Mattelart, A. and H. Schmucler (1985) *Communication and Information Technologies: Freedom of Choice for Latin America?* Norwood, NJ: Ablex Publishing Corporation.

McGowan, C. and R. Pessanha (1998) *The Brazilian Sound: Samba, Bossa Nova, and the Popular Music of Brazil*. Philadelphia, PA: Temple University Press.

Miller, T. (2001) *Global Hollywood*. London: British Film Institute.

Miller, T., N. Govil et al. (2005) *Global Hollywood 2*. London: Pearson Longman.

Morley, D. (1980) *The Nationwide Audience: Structure and Decoding*. London: British Film Institute.

Morley, D. (1992) *Television, Audiences and Cultural Spaces*. New York, NY: Routledge.

Morris, N. and S. Waisbord (eds) (2001) *Media and Globalization: Why the State Matters*. Lanham, MD: Rowman & Littlefield.

Naficy, H. (1993) *The Making of Exile Cultures: Iranian Television in Los Angeles*. Minneapolis, MN: University of Minnesota Press.

Nordenstreng, K. and T. Varis (1974) *Television Traffic—A One-Way Street*. Paris: UNESCO.

O'Connor, A. (2002) "Local scenes and dangerous crossroads: Punk and theories of cultural hybridity," *Popular Music*, 21: 225–236.

Ogan, C. and M. Milikowski (1998) *Boundaries Crossed and Maintained: Turkish Migrants' Culture, Religion and Use of Satellite Television*. Jerusalem: ICA.

Orozco Gómez, G. (1996) "La TV a fin de milenio: Un medio poderoso, cuyo límite es la audiencia, in G. Orozco Gómez (ed) *Miradas latinoamericanas a la televisión*. México: Universidad Iberoamericana.

Papastergiadis, N. (2000) *The Turbulence of Migration: Globalization, Deterritorialization, and Hybridity*. Malden, MA: Blackwell.

Pieterse, J. N. (2004) *Globalization and Culture*. Lanham, MA: Rowman & Littlefield Publishers.

Rantanen, T. (2005) *The Media and Globalization*. London: Sage Publications.

Rojas, V. (2001) "Latinas in El Show de Cristina: Notes on representation within a framework of sexual labor", ICA Conference.

Schiller, H. I. (1981) *Who Knows: Information in the Age of the Fortune 500*. Norwood, NJ: Ablex Publishing Corporation.

Sinclair, J. (1999) *Latin American Television: A Global View*. New York, NY: Oxford University Press.

Sinclair, J. S., E. Jacka et al. (1996) "Peripheral vision," in J. S. Sinclair, E. Jacka et al (eds) *New Patterns in Global Television*. New York, NY: Oxford University Press, 1–15.

Søndergaard, H. (2003) "Television and national identity in Denmark: The return of the nation," in S. Hjarvard (ed) *Media in a Globalized Society*. Copenhagen: University of Copenhagen Press, 15–53.

Sparks, C. (1998) "Is there a global public sphere?" in D. K. Thussu (ed) *Electronic Empires: Global Media and Local Resistance*. London: Arnold.

Sparks, C. (2005) "The problem of globalization," *Global Media and Communication*, 1(1): 20-23.

Straubhaar, J. (1991) "Beyond media imperialism: Asymmetrical interdependence and cultural proximity," *Critical Studies in Mass Communication*, 8: 39–59.

Straubhaar, J. (2005) "(Re)Asserting national television and national identity against the global, regional and local levels of world television," in G. Durham and D. Kellner (eds) *Media and Cultural Studies: Keyworks*, revised edition. New York, NY: Blackwell, 681–702.

Straubhaar, J. (2007) *World Television: From Global to Local*. Newbury Park, CA: Sage Publications.

Telles, E. E. (2004) *Race in Another America: The Significance of Skin Color in Brazil*. Princeton, NJ: Princeton University Press.

Thussu, D. K. (ed) (2007) *Media on the Move: Global Flow and Contra-Flow*. New York, NY: Routledge.

Vargas Llosa, M. (1981) *La guerra del fin del mundo*. Barcelona: Editorial Seix Barral.

Vianna, H. (1999) *The Mystery of Samba*. Chapel Hill, NC: University of North Carolina Press.

Weber, E. J. (1976) *Peasants into Frenchmen: The Modernization of Rural France, 1870–1914*. Stanford, CA: Stanford University Press.

Chapter 2

Mapping Arab Television

Structures, Sites, Genres, Flows, and Politics

Marwan M. Kraidy

During the past 20 years, the Arab world—consisting of 22 Arab-speaking countries in North Africa and Western Asia—has witnessed a proliferation of satellite television channels of such proportions that some observers have labeled it a "revolution" (Information Revolution in the Arab World 1994). The process started with the establishment in London in 1990 of the Saudi-owned Middle East Broadcasting Center. It made a significant qualitative leap in 1996 with the rise of the news network al-Jazeera from Doha and the beginning of satellite transmission by the Lebanese entertainment channels LBC-Sat and Future TV. Finally, the emergence of the al-Arabiya network to rival al-Jazeera, the mushrooming of Iraqi channels in wartime anarchy, and the proliferation of specialized music, women's business, and religious channels since 2003 have established the Arab world as the scene of a vibrant indigenous media industry. Questions about the aptness of the label "revolution" aside, between 1990 and 2010 the Arab world has gone from zero to approximately 500 Arabic-language satellite channels (Kraidy and Khalil 2009).

This staggering growth in the number of media outlets coincided with a radical qualitative transformation. As Arab television evolved from separate state-owned, nationally based, terrestrial broadcasting services to a commercial, transnational satellite industry, new links developed between institutions, governments, and audiences, enabled by the convergence of satellite, Internet, and mobile media across a multinational area with approximately 300 million inhabitants. This has led to an efflorescence of television genres, to the de-territorialization of media production, to a continuing regionalization of Arab television, and the rise of contentious politics as a recurrent feature of this dynamic transnational media sphere connected by the hardware of geostationary satellites and the software of the Arabic language.

This chapter offers a preliminary exploration of the structures and flows that characterize Arab television and maps them onto national, regional, and global scales, taking into account industry structures, programming flows, and the transnational contentious politics that flows of content generate in the Arab world. In addition to brief analytical descriptions of developments in drama, comedy, and other genres, I draw on my recent work on the Arab reality television polemics (Kraidy 2009) to build conceptual and empirical bases in order to understand the industry. The pan-Arab reality television polemics of the last decade provide a heuristic opportunity to revisit some of the core topics, theories, and methodologies in global communication studies, including, first, changing relationships between the local and the global and, second, a shifting political economy in which interactive features using mobile telephony and the Internet are embedded in television programs.

In addition, with an eye toward possible future direction that global communication studies may take as a field, this chapter discusses, third, the often-neglected role of contentious politics in shaping global media industries and flows. Finally, I revisit notions of proximity and other cultural, social, and historical identities/affinities in order to re-elaborate them as constitutive of, fourth, a theory of multiple modernities in which media institutions play a central role. The chapter concludes with a series of concrete recommendations for developing new agendas in global communication studies, emphasizing the rising importance of regionalization, contentious politics, and a renewed focus on the issue of modernity.

Anatomy of a De-Territorialized Industry

During the last two decades, significant changes have occurred in the sites, structures, genres, and flows of Arab television. The overall structure of Arab television (its sites of production, nodes of distribution, and target audiences) has entirely changed. Governments have established media cities in Dubai, Amman, Damascus, and Cairo, while Beirut remains a *de facto* media center *in spite of*, not as a result of, media policies. This concentration of capital, human resources, and media activity in sites often kept at an arm's length from direct government intervention has changed the nature of media exchanges from cooperation between states, under the aegis of the Arab States Broadcasting Union, or, more broadly, the Arab League and its sub-agencies to joint ventures, sub-contracting, and the movement of creative labor between media companies located in Beirut, Cairo, and Dubai. At the same time, rising multiplatform conglomerates such as MBC Group, Rotana-LBC, and others have divided their activities between these three cities and others. For instance, MBC's headquarters are located in Dubai, but most of its productions have come out of Beirut and Cairo (Kraidy and Khalil 2009).

There are also important changes in the roles of established production centers and the emergence of new loci of television production. Before the advent of satellite television, for example, Egypt was *the* main center for drama production, and virtually all state-owned Arab channels programmed Egyptian drama. In the satellite age, Egypt's longtime dominance in drama production has eroded, losing significant ground to Syrian drama, which has emerged as a potent challenger to Egypt's leadership in Arab television drama after investment legislation in the 1990s allowed private capital—a significant portion of it coming from the oil-rich Gulf states—into Syrian drama production (Kraidy 2006b), which turned members of the Syrian regime, most notably the son of erstwhile Vice President Abdul Halim Khaddam, into media moguls. However, the economic underpinnings of Syrian drama are now vulnerable to the development of indigenous drama production in the Gulf, especially in Kuwait and Saudi Arabia, with various degrees of popular success, which risks siphoning away production capital from Syrian drama. At the same time, with drama production increasingly decoupled from state institutions and subjected to market rules, competition between Egyptian and Syrian drama has also led to a variety of co-productions, bi-national casts, and the mobility of creative labor, a notable trend being major Syrian actors taking lead roles in Egyptian drama (Kraidy and Khalil 2009).

Drama increasingly shares primetime with other television genres. With the growing commercialization of Arab television, global customary staples of advertising-supported genres have emerged in force, including game and variety shows, music videos, reality television, and niche programs targeting youth, women, and religious people. In this context, some nation states (and media cities) have emerged as a locus for the production of particular genres. Music videos are chiefly the province of Egypt and Lebanon, perhaps the two most socially liberal Arab countries. Comedy, a notoriously bad border-crosser, remains stubbornly national. A wide gamut of religious channels ranging from Salafi-Wahhabi to Light-Islamic appear to be largely the product of flows of money and expertise between Egypt and Saudi Arabia (and, to some extent, Jordan), while the Levant is the scene of religiously themed politically radical channels such as Hezbollah's al-Manar and Hamas' al-Aqsa in Lebanon and Palestine.

Programming and distribution have also changed dramatically as media executives have to take into account multiple cross-cutting vectors including social class, gender, age, religious affiliation, sub-regional differences, national proclivities, and intra-national specificities. Various combinations of these factors fragment audiences and shape production, programming, and advertising strategies. For example, because of large concentrations of wealthy viewers in the Gulf countries, many productions target them at the expense of other markets, especially the neglected North African viewers. Until recently, Arab television industries consisted of Gulf capital, Levantine and Egyptian production, and Gulf audiences. Lately, pan-Arab channels such as LBC and MBC have attempted to lure viewers from the Maghreb, while satellite channels in Algeria, Morocco, and Tunisia have developed Maghreb-focused programs such as *Star Academy Maghreb* and others.

Arab television today is therefore a de-territorialized industry whose production, distribution, and consumption operate across cultural, national, and sub-regional boundaries in a region of the world defined primarily, for the sake of the media industries, by the shared Arabic language. Production traffic is for the most part a translocal—or, more aptly, transurban—phenomenon between Beirut, Cairo, and Dubai, connected by a mobile creative class and by capital from the oil-rich Gulf states. Distribution and transmission are transnational, with audiences cobbled together based on sub-regional identities (i.e., Levant, Gulf, and Maghreb, or other demographic and lifestyle characteristics). Industry practices, production locations, distribution patterns, and program flows in today's Arab television, then, clearly reflect a changing relation between the local (city), the national, and the regional (pan-Arab). Recently, global players have entered the industry; as a result the political economy of Arab television is experiencing an increased integration in the global media market.

A Shifting Political Economy

The Entrance of Global Players

On September 28, 2009, the Lebanese leftist daily newspaper *al-Akhbar* broke a story about massive layoffs in the venerable right-of-center Beirut daily *Annahar* after a Booz Allen Hamilton study commissioned by the newspaper's owners recommended a radical restructuring. Headlined "*Annahar*, End of an Era: More than 50 Editors and Employees Laid Off" (Ayoub 2009) the story in which Beirut's newest daily newspaper scooped the Arab media establishment in revealing a financial crisis at Beirut's oldest daily was the first in a series of articles, columns, and letters to the editor about a series of layoffs and structural adjustments touching various Lebanese media outlets. On October 1, *al-Akhbar* ran a story entitled "*Annahar*'s Influenza Infects Media Institutions" (Haddad 2009). In addition to *Annahar*, most strongly affected were the Lebanese Broadcasting Corporation (LBC) and Murr Television (MTV).

What *Annahar*, LBC, and MTV have in common is that they are each partly owned and financed by the Saudi royal prince, global investor, and media mogul al-Waleed bin Talal, whose global empire of companies and investments had taken a serious hit in the global economic recession, and whose Rotana media conglomerate was announcing mergers of several of its channels after having moved its production headquarters from Beirut to Egypt in a cost-cutting measure. At the same time, al-Waleed's stake in LBC's satellite channel had grown to approximately 90 percent, making it a *de facto* Saudi channel. In turn, stories swirled in the Arab press that Rupert Murdoch, a friend and business partner of al-Waleed—the prince owns a substantial stake in Murdoch's News Corporation and is rumored to have helped the Australian tycoon at times of vulnerability—was interested in purchasing a high number of shares in Rotana. As of early February 2010, both the US and Arab press carried reports that discussions between the two sides were, in fact, under way (Muhammad 2010; Sergie and Critchlow 2010).

Emerging links between LBC, Rotana, and News Corporation constitute the first example of an initially national player, LBC, which began as a platform for a Christian-Lebanese nationalist party, which then became a national–regional television player as various Saudi moguls invested in it, finally restructuring its operations to make its controlling regional music and television conglomerate (Rotana) a better prospect for investment by a global corporation (News Corporation). Nonetheless, these developments reflect growing transnational and translocal links, partnerships, joint ventures, and mergers and acquisitions in the Saudi-dominated pan-Arab media landscape, which, in turn, is beginning to attract the attention of global media tycoons.

In this context, the recent Lebanese media industry crisis can be understood as a side-effect of what elsewhere I have called the "Saudi-Lebanese" connection, in which Saudi capital and media institutions employ a disproportionately large number of Lebanese, who shape the aesthetics and branding of entertainment shows, most directly reality television. In the small size of the country from which they hail, their forced displacement due to political upheaval, their instrumental role in the advertising industries, their proximity to dominant media markets, their know-how, and the outsized influence of the Lebanese on the pan-Arab television industry bear striking resemblances to the influence of Cubans in the Latin American television industry (Rivero 2009), a comparison that could be pursued further, perhaps through the prism of the trajectories of creative migration described by Curtin (2007).

From Ratings to "Engagement"

There is another shift in the political economy of Arab television, and that is a new business model that LBC pioneered in the context of its reality television productions. The highly visible wave of reality television shows that have filled Arab primetime since 2004 reflects the emergence of a new political economy based on pay-by-the-click interactive features relying on mobile telephony and the Internet. This enables the industry to circumvent the lack of reliable audience research which has kept advertising spending in the Middle East and North Africa at US$2 billion per annum, the second lowest in the world after sub-Saharan Africa, in spite of the immense oil wealth that entered the region in the last decade (*Campaign Middle East* 2006a). As a result, Arab media companies have developed new revenue streams by harnessing interactive features in Short Messaging System (SMS), Multimedia Messaging System (MMS), and Interactive Video Network (IVN). The use of interactivity has changed the debate about audience measuring techniques from a concern for quantitative measures provided by people-meters installed in television sets, to an emerging interest in qualitative features of media consumption under the rubric of "engagement." Hence the argument by an executive at Omnicom Media Group (OMD) Middle East that:

> Everyone will tell you that people meters are crucial, and that's all good, but the issue in my opinion is that TV meters are passé, an old story ... We should go beyond TV meters and I think in this region we have the opportunity to go to the next stage. *It's about doing engagement*

> *studies, how programmes are engaging with viewers—not just how long they're spending watching them.*
>
> *Campaign Middle East* (2006b, p6; emphasis added)

In reality television, the Arab television industry has found an opportunity to test new business models based on interactivity and made possible by media convergence, enabling the rise of what I elsewhere called *hypermedia space* (Kraidy 2006a) after Deibert's discussion of hypermedia (Deibert 1997), which integrates television with the Internet and mobile phones. Reality television is particularly suited for hypermedia space because the reality genre is premised on the viewers nominating and voting for contestants via text-messaging or websites. Such practices give television channels demographic information useful for targeting. It is therefore not surprising that during the 2007/2008 season, 16 out of 26 (61.5 percent) Arab reality shows offered viewer interactivity (Snobar 2008). While giving commercial media corporations additional revenue streams, the changing political economy of television has also contributed to the rise of mediated contentious politics in the Arab world, to which we turn next.

The Role of Contentious Politics

In this section, I would like to make an argument to include contentious politics as an important force shaping the development of media industries. Though this may be peculiar to the Arab world, it is a sufficiently influential factor to warrant discussion that is not restricted to a "culture wars" framework, but rather takes into account that contentious politics can influence the political economy of media. Historically, military conflict and propaganda wars have shaped the growth of Arab media. In reaction to Nasser's radio attacks during the late 1950s, the Saudi royal family developed a television service, in the meantime sacking numerous Egyptian journalists from Saudi media institutions when the two countries were on opposite sides of the Yemen war. In effect, Nasser's anti-Saudi broadcasts led to the emergence of Saudi Arabia's national broadcasting structure. Similarly, al-Jazeera's ratings skyrocketed during the Anglo-American invasion of Afghanistan in 2001, and its criticism of the Saudi regime led to the 2003 launch of al-Arabiya. During rhetorical wars between Saudi Arabia and Qatar, Lebanon and Syria, Saudi Arabia and Syria, Saudi Arabia and Iran, and Egypt against Hezbollah, Arab governments and media tycoons have enacted reforms, restructured institutions, and developed partnerships that have changed the structure of the media industry in the region.

The reality TV controversies have since 2003 captured the transformation of a protracted Arab malaise into an acute existential crisis precipitated by the attacks of 9/11, the invasion of Iraq, bloodshed throughout the region, and historic Arab realignments along the US–Iran contest for hegemony over the Middle East. This has been a period of collective soul-searching during which Arabs have debated norms governing their public life and grappled with shifting identities and changing values. The process is public and contentious: it arises in homes, coffeehouses, and intellectual *fora*, and unfolds on television screens, newspaper pages, the Internet, and even mobile phones. By stirring up a volatile mix of politics, religion, business, and sexuality, reality TV has compelled more Arabs to join heated deliberations over longstanding and ponderous issues such as Western intervention, gender roles, political participation, and changing norms, values, and identities.

Since 2004, reality TV has entered the fray of Arab politics, getting embroiled in legislative debates, contested elections, diplomatic incidents, and executive–legislative tensions. It started in Bahrain, where a scandal surrounding *al-Ra'is*, the local version of the global hit *Big Brother*, led to heated discussions in the national assembly that elicited two rival understandings of "national reputation" whose hitherto established meaning as conformity to Islamic values was now contested by an emerging, perhaps best understood as neoliberal, definition of national reputation as fitness for foreign investment. The Saudi outcry over reality television was even more impassioned: After *Star Academy* attracted record ratings, Saudi radicals, conservatives, and liberals recruited reality television as a particularly controversial trope in their culture wars, prompting a supreme religio-juridical council to issue a *fatwa* warning against the show's many perceived dangers. In neighboring Kuwait, Islamist legislators forced a minister of information to resign for allowing *Superstar* contestants to organize a concert as they fought against women's political enfranchisement (see Kraidy 2009).

The reality television wars illustrate how contentious politics revolving around media content can be decisive contributors to structural changes in the industry. As I discuss in detail in *Reality Television and Arab Politics*, Algerian president Abdelaziz Bouteflíqa faced intense criticism from the Islamist-dominated parliament for allowing the Lebanese production of *Star Academy* to be relayed on Algerian state television. Because he needed the support of Islamist legislators on other issues, Bouteflíqa interrupted the broadcasts despite the show's enormous popularity among young people. Algerian state television then replaced it with a local and more conservative version of *Star Academy*. One year later, that show was canned in favor of a Quranic recitation program, sealing the Islamists' victory over the executive and highlighting Bouteflíqa's desperate need for religious legitimacy. The situation in Algeria, in addition to similar developments in Tunisia, where reality television shows were under attack by conservatives but enjoyed significant popularity across audience segments, led Nasma TV, a privately owned Tunisian outlet, to develop *Star Academy Maghreb*, targeting Algerian, Tunisian, and Moroccan viewers, contributing to turning the Maghreb into a supranational and sub-regional media working space for advertisers and programmers alike. This, in addition to other Maghreb themed shows, in turn undermined efforts by LBC and MBC to win over North African Arab-speaking audiences, hence solidifying segmentation along the lines of the Levant, Gulf, and Maghreb, creating ready-made circuits and audiences for emerging channels (Kraidy 2009).

Critics of reality television in the Arab world were most incensed by reality shows because they perceived them to be a direct affront to Muslim faith and Arab dignity because of *reality TV's claims to represent reality.* Reality TV shows feature amateurs and, in the case of *Star Academy*, beamed in live and continuous satcast or webcast flows, 24/7. This makes their content unpredictable to their audience and justifies their claim to the real. In the West, some surveys have shown that most viewers are skeptical of reality TV's claim to be real (Hill 2005), so the genre's claim to "the real" can be dismissed. In the Arab world, representational claims are contentious not only because viewers mistrust reality TV's pretense to be real, but also because of longstanding grievances that circulate in popular discourse in the Middle East about Western misrepresentations of Arabs and Muslims, leading to an acute sensitivity toward representations of Arab and Muslim issues. Caught in a global web of representation that often disparages them, Arabs cannot afford to dismiss television as "just TV." Shortly we will see that Muslim clerics vigorously contested reality TV's claim to represent reality because the lifestyle, behavior, and values manifest on *al-Ra'is* and *Star Academy* are anathema to these clerics' vision of a virtuous society. From their perspective, they were resisting the imposition of a foreign reality on their world (see also Hammoud 2008). Reality TV's declarations about reality are antithetical to prevailing forms of social authority and cultural representation. The contest around defining social reality has a direct bearing on how the notion of modernity can be conceptualized as a heuristic device building on existing concepts such as cultural proximity while suggesting new avenues of inquiry in global communication studies.

Multiple Proximities, Multiple Modernities

The reality TV controversies highlight contrary views of the Arab experience with *al-hadatha*: modernity. What it means to "be modern" in the Arab context has been vigorously contested since the 1850s, but the debate took on a renewed poignancy in the 1990s with the rise of pan-Arab commercial television and its growing linkages to the global media market. Despite countless debates over its definition, it is generally agreed that modernity entails self-criticism through which societies constantly renew themselves. As the sociologist Anthony Giddens (1990, p76) put it, modernity "effectively involves the institutionalization of doubt." On the surface, modernity presents the Arab world with a challenge. Can Arab societies, where Islam permeates social, cultural, and political life, follow the modern path of decreasing the role of religion in public life? The answer may be no if we follow Giddens, who writes that modernity is "manifestly incompatible with religion as a pervasive influence upon day-to-day life" (Giddens 1990, p109). But the answer could be yes according to scholars who have argued that modernity has theological origins, that religious speech enabled early European modernity, that there are different kinds of modernity, and that, more specifically, media usage helped Arab and Muslim communities to accommodate modernity without forsaking their heritage (see, among others, al-Azmeh 1993; Blanks 2001; Cole 1999; Deeb 2006). The commotion over reality TV played up tensions associated with modernity—the relation between religion and the state, the development of representative forms of governance, the rise of nationalism, and conflict over gender roles.

Debates over the meaning of modernity are heated in the non-West because "modernity" conjures up social progress, economic growth, individual emancipation, or cultural modernism; or, alternatively, cultural decline, loss of authenticity, and economic dependency. Complicating these discussions is the widespread belief that modernity is incapable of shedding its Western ethos. Nonetheless, if modernity involves what the philosopher Charles Taylor (1999, p168) called "the coming to be of new kinds of public space," then impassioned debates about reality TV constitute crucial episodes of Arab engagement with modernity. It would be too easy to conceive of the Arab reality TV debate as a clash between tradition and modernity. Rather, under the irresistible pull of Western modernity, societies search for what Taylor dubbed "creative adaptation" (Taylor 1999, p163) to adopt constituents of modernity. Though broadly sympathetic to the multiple modernities literature (Gaonkar 1999; Mitchell 2000; Hervieu-Léger 2003; Kaya 2004), like Mitchell (2000), I am aware of modernity's insistent claim to universality and its unavoidable linkages to the West. Through a critical engagement with theories of modernity, this book aims to explicate anew how the struggle between rival versions of modernity—what the philosopher Leszek Kolakowski (1990) called modernity's "endless trial"—unfolds.

The controversy around reality television is fuelled by the cultural hybridity embedded in those programs. As local adaptations of global formats, Arab reality shows are hybrid texts that mix foreign and local cultural sensibilities (see Kraidy 2005). Reflecting the lifestyle of the socially liberal, highly mobile, and Westernized creative class who directs and produces them, Arab reality TV shows clash with conservative mores in the region and more specifically in Arabian Gulf societies, whose moguls and princes paradoxically invest in reality TV productions and whose well-to-do viewers are prime targets of the corporations who advertise on reality TV programs.

Writing about Latin America, Néstor García-Canclini (1998, p2) pithily remarked that "the uncertainty about the meaning and value of modernity derives not only from what separates nations, ethnic groups and classes, but also from the socio-cultural hybrids in which the traditional and the modern are mixed." Arab poetry, art, literature, and cuisine are replete with various hybridities, but reality TV emerged as a highly visible, publicly contentious, cultural hybrid. Arab reality TV is so intensely controversial because it violates boundaries of identity and authenticity at a time when these boundaries have been hardened by widespread violence in the Middle East and various global controversies over Islam. For some opponents, *al-Ra'is* and *Star Academy* were links in a chain of horrors—Abu Ghrayb, Guantanamo, the infamous Danish cartoons, and others befalling Arabs and Muslims. Nonetheless, the arguments about reality television focused not on a wholesale rejection of Western or foreign values, norms, and styles, but on ways in which the harmful dimensions of Western modernity could be kept out or at least mitigated, while the beneficial aspects could be appropriated.

This search for a socially accessible, religiously acceptable, and culturally resonant modernity can be glimpsed in other developments in Arab media and culture. Over the last three years Turkish drama serials, dubbed in the Levantine Arabic dialect, have taken the Arab world by storm, causing controversy and moral panics in the Gulf countries, especially Saudi Arabia. There are some antecedents of non-Arab shows achieving great popularity in the region. In some of the Gulf countries facing India, especially Oman, Bollywood films have historically done very well. The "Turkish Serials Phenomenon" also brings to mind the popularity among Arab audiences of Latin American *telenovelas* from the mid-1980s to the late 1990s.

At the same time, the politico-economic backdrop and various proximities between Turkey and the Arab world—literal (i.e., geographical proximity) but also religious, social, culinary, and historical in that they were both integrated in the Ottoman Empire—make the reception of Turkish serials qualitatively different. The Arab press has been abuzz with stories about the rise of Arab tourism and investment in Turkey as a result of these broadcasts. Elsewhere (Kraidy 2005), I attempted to theorize the notion that there is a "global south aesthetic" to some television genres that make them travel with relative ease on South-to-South cultural circuits (see LaPastina and Straubhaar, 2005, for Brazilian–Mexican and Brazilian–Italian connections). The popularity of Turkish drama with Arab audience offers additional intriguing theoretical possibilities to explore proximities in connection to the literature on multiple modernities (for related approach in East Asia, see Iwabuchi 1997).

The interplay of multiple modernities and multiple proximities is driven by changing inter-scalar relations—between local, national, regional–pan-Arab, and global—in the structure of Arab television. The ways in which the national and regional scales relate to each other have changed. For more than two decades, national channels broadcast terrestrially while pan-Arab channels, for obvious reasons, reached viewers via satellite. Recent trends reflect a growing use of satcasting by channels with an essentially national scope. At the same time, simultaneous processes of consolidation and fragmentation can be observed, with the advent of multiplatform conglomerates coinciding with a marked growth in increasingly specialized channels; a process of differentiation is also unfolding *within* niche markets, such as business television, with channels emerging that specialize in finance, real estate, etc. (see Kraidy and Khalil 2009). As a result, special interest audiences are increasingly defined across scales.

Linguistic and cultural forces remain powerful shapers of the media industries, with differences between nation states emerging at the expense of the pan-Arab identification that some analysts of Arab media observed. For example, newscasts tend to be presented in *fusha*, or modern standard Arabic, but talk shows, *musalsalat* (drama series), and variety programs expose Arab viewers to a variety of accents, styles, and cultures. The Syrian *musalsal Bab-al-Hara* (*The Neighborhood's Gate*) have also rekindled interest in hyper-local identities, in this case that of Damascus. The national provenance of television workers adds a layer of scalar complexity: the majority hails from Egypt, Jordan, Lebanon, and Syria, including many Palestinians from these countries; a smaller number hails from the Gulf thanks to policies aiming to increase the number of national television workers; and very few are from the Maghreb countries in North Africa. Algeria, Libya, Morocco, and Tunisia remain relatively marginal in the pan-Arab television industries as sites of production, targets of programming, and advertising markets, though some media corporations based in the Levant and the Gulf have given more attention to the Maghreb countries.

The changing relationship between national and pan-Arab scales has driven a wave of national media reforms through which nation states have attempted to project their political and economic agendas with a distinct "national" voice. Saudi Arabia, Bahrain, and Syria, for instance, followed a top-down approach to revamping their national television systems, with the state remaining in full control. Even when countries such as Egypt, Jordan, and Syria allowed privately owned, regime-friendly private satcasters, many of these companies have faced hurdles establishing themselves as a result of constant state interference and policy changes. Other Arab states followed a path of liberalization and allowed privately owned satellite channels to operate on their territory. This should be understood in the context of regimes' desire to attract viewers to politically friendly channels. Privately owned satellite channels such as al-Rai in Kuwait, Dream TV in Egypt, and Hannibal TV in Tunisia have increasingly targeted national viewers. In the case of Egypt and some others, the heavy hand of the state, in tandem with business moguls primarily interested in the domestic market, has prevented the development of privately owned Egyptian networks with significant pan-Arab viewership. In the Maghreb countries—Morocco, Tunisia, Algeria—neglect by satellite broadcasters based in the Levant and the Gulf has led to attempts by some state broadcasters such as Algeria's to offer exciting but socially non-controversial entertainment television, while pushing others such as Tunisia's along a path like Kuwait's, where national channels have used satcasting to circumvent restrictive regulatory measures.

Scalar changes can be also observed in the unmistakable trend toward market fragmentation in the Arab television industries, as well as demographic differentiation, niche programs, and specialized channels. Gulf viewers are known to have more disposable income than other Arabs, so the priority of most Arab television institutions is to build market share in the Gulf states.

Among those, Saudi Arabia is by far the most important market, not only because of demographic size and income levels, but also because success with Saudi viewers often opens doors to Saudi petro-dollars for television production. Not only do Saudi investments fuel production of drama in Syria, reality television in Lebanon, and music videos in Egypt, but Saudi money sustains the leading production centers in the Arab world, including media cities in Jordan and Dubai. Increasingly, Arab producers and media owners pursue Saudi financing by tailoring their productions to the proclivities of Saudi viewers and media moguls alike. On a grand scale, the influence of Saudi money, taste, and production values is increasingly visible throughout Arab television industries.

The consolidation of Arab television industries into pan-Arab multiplatform television operators, which include the MBC Group, the al-Jazeera network, and the LBC-al-Hayat-Rotana conglomerates, also drives market fragmentation. Channels in these conglomerates provide broad news and entertainment programming while offering a wide gamut of specialized programs ranging from Arabic pop music to shows about personal investments, to sports, US action movies, and special programs for political or religious occasions. The rise of these groups has a twofold impact. On the one hand, overall it has boosted local production, enhanced program quality across genres and interests (which nonetheless remain uneven), and increased advertising revenues. On the other hand, consolidation has fomented a commercial race that risks intensifying audience flight to the few dominant conglomerates, thus weakening production quality and undermining national production diversity.

Conclusion

The contemporary pan-Arab television industry is dynamic, pluralistic, and ever-changing. Driven by geopolitical instability, enormous profits generated by energy prices, real estate speculation, and the economic problems affecting parts of the Arab world—most spectacularly Dubai—and a host of religious and social factors, the Arab television industries reflect a complex mix of business, culture, and politics. The Arab world's geopolitical and economic importance rising in proportion to the world's economy needs for energy and Arab media are poised to continue growing and experiencing a process of consolidation that, as the Newscorp–Rotana pending deal reflects, increasingly integrates Arab media within the global market. The historically unique speed and scale of development in the Arab media industry, and the complexity inherent in the myriad military, economic, political, ideological, cultural, and religious forces impinging on Arab media, make that scene a heuristic space from which to consider items for new agendas in global communication studies.

Many lessons can be drawn from this chapter, but three stand out. First, *the impact of globalization on the media industries is perhaps best understood at the regional-transnational level, rather than the national.* Though this has become conventional wisdom, the challenge is that by shifting the unit of analysis away from nation states and national media industries, the object of research becomes fluid and there is an absence (or, at the very least, a scarcity) of regulation and policies. The global–local dyad that has preoccupied international communication research needs to be re-conceptualized. Our analysis must focus on regionally bounded, globally linked, locally inflected interactions between nation states and their policies, media tycoons, social actors, and media markets (see Curtin 2007; Straubhaar 2007; Kraidy 2009). Though several people have worked on this, we need to *investigate and theorize regionalization more systematically* to take into account the inter-scalar complexity of global communication.

A second lesson could be a starting point to re-examine some of the longstanding assumptions guiding international communication research. *Multiple flows and proximities operating across fluid structures and shifting scales have a high level of built-in contingency.* Trying to explain why Mexican *telenovelas* or Turkish soap operas were successful with Arab audiences (as opposed to, say, Iranian or Japanese productions) leads us to a complicated transcultural political economy where structures of distribution, programming practices, and various cultural proximities interact to produce many possible outcomes. This is an opportunity to develop macro-theories of multiple modernities in global media studies, a development that promises to deliver a more flexible and sounder framework than the local–global dichotomy. Though a multiple-modernities approach recognizes that "the West" is unavoidable structurally and normatively, it nonetheless does not re-center the West by conflating it with the global. I would argue that the defining and distinct contribution of a multiple-modernities approach would be to emphasize South-to-South or local-to-local connections (the terminology is problematic, and developing a more rigorous lexicon should be part of the effort).

Finally, global communication scholars would benefit from considering the kinds of mediated activities of political and social contestation that have become known in the social movements literature as "contentious politics." Indeed, I hope to have shown that these kinds of contestations have an impact upon media structures. *Contentious politics contribute to shaping the changing structure of the Arab television industries and vice-versa.* Important episodes include war, rhetorical wars among Arab regimes, as well as various episodes of contention under the rubric of culture wars, moral panics, and assorted scandals. Integrating these issues would enrich our understanding of the media industries. At the same time, this calls for a re-examination of the media events to account for the changing media environment, including emerging media—mobile, social networking, etc. (see Kraidy, 2009, and Punathambekar's notion of "mobile publics," 2010). This could also be conceived as another way to understand what Curtin (2007) calls "forces of socio-cultural variation" (2007), though the contentious politics formation in the literature on social movements remains unexplored in communication and media studies.

References

al-Azmeh, A. (1993) *Islams and Modernities*, London: Verso.

Al-Sharyan, D. (2009) "The sun of Lebanon's *annahar*," *Al-Hayat*, 30 September.

Ayoub, S. (2009) "*Annahar*, end of an era: More than 50 editors and employees laid off," *Al-Akhbar*, 28 September.

Blanks, J. (2001) *Mullahs on the Mainframe*. Chicago, IL: University of Chicago Press.

Campaign Middle East (2006a) "Choueiry: it's time to raise ad spending," *Campaign Middle East*, 3 December.

Campaign Middle East (2006b) "We must go past TV Meters, says OMG boss," *Campaign Middle East*, 24 September, p6.

Cole, J. (1999) *Modernity at the Millennium*. New York, NY: Columbia University Press.

Curtin, M. (2007) *Playing to the World's Biggest Audience: The Globalization of Chinese Film and Television*. Berkeley, CA: University of California Press.

Deeb, L. (2006) *An Enchanted Modern*. Princeton, NJ: Princeton University Press.

Deibert, R. (1997) *Parchment, Printing, and Hypermedia*. New York, NY: Columbia University Press.

Gaonkar, D. P. (1999) "On alternative modernities," *Public Culture*, 11(1): 1–18.

García-Canclini, N. (1998) *Consumers and Citizens: Multicultural Conflicts in Globalization*. Minneapolis, MN: University of Minnesota Press.

Giddens, A. (1990) *The Consequences of Modernity*. Palo Alto, CA: Stanford University Press.

Haddad, L. (2009) "*Annahar*'s influenza infects media institutions," *Al-Akhbar*, 1 October.

Hammoud, 'A. (2008) *Reality Television: Humans in the Cage of the Image*. Beirut: Dar al-Hadi.

Hervieu-Léger, D. (2003) "Pour une sociologie des 'modernités religieuses multiples': Une autre approche de la 'religion invisible' des sociétés européennes," *Social Compass*, 50(3): 287–295.

Hill, A. (2005) *Reality TV: Audiences and Popular Factual Television*. London: Routledge.

Information Revolution in the Arab World (1994) Symposium, Center for Contemporary Arab Studies, Georgetown University, Washington, DC, March.

Iwabuchi, K. (1997) "The sweet scent of Asian modernity: The Japanese presence in the Asian audiovisual market," Paper presented at the Fifth International Symposium on Film, Television and Video, Media Globalization in the Asia-Pacific Region, Taipeo.

Iwabuchi, K. (2002) *Recentering Globalization: Popular Culture and Japanese Transnationalism*. Durham, NC: Duke University Press.

Kaya, I. (2004) "Modernity, openness, interpretation: A perspective on multiple modernities," *Social Science Information*, 43(1): 35–47.

Kolakowski, L. (1990) *Modernity on Endless Trial*. S. Czerniawski, W. Freis, and A. Kolakowska (trans.). Chicago, IL: University of Chicago Press.

Kraidy, M. M. (2005) *Hybridity, or the Cultural Logic of Globalization*. Philadelphia, PA: Temple University Press.

Kraidy, M. M. (2006a) "Governance and hypermedia in Saudi Arabia," *First Monday*, 11(9), http://firstmonday.org/issues/special11_9/kraidy/index.html.

Kraidy, M. M. (2006b) "Syria: Media reform and its limitations," *Arab Reform Bulletin*, 4(3), May, Carnegie Endowment for International Peace, Washington, DC, http://www.carnegieendowment.org/publications/index.cfm?fa=view&id=18341#media.

Kraidy, M. M. (2009) *Reality Television and Arab Politics: Contention in Public Life*. Cambridge and New York: Cambridge University Press.

Kraidy, M. M. and J. Khalil (2007) "The Middle East: Transnational Arab television," in L. Artz and Y. Kamalipour (eds) *The Media Globe: Trends in International Mass Media*. Lanham, MD: Rowman & Littlefield, 79–98.

Kraidy, M. M. and J. F. Khalil (2009) *Arab Television Industries*. London and New York: British Film Institute/Palgrave Macmillan.

LaPastina, A. and J. Straubhaar (2005) "Multiple proximities between television genres and audiences: The schism between *telenovelas*' global distribution and local consumption," *Gazette*, 67: 271–288.

Mitchell, T. (2000) (ed) *Questions of Modernity*. Minneapolis, MN: University of Minnesota Press.

Muhammad, H. (2010) "The coming alliance between Murdoch and al-Waleed: A new Arab media phase?," *Al-Quds al-Arabi*, 31 January.

Punathambekar, A. (2010) "*Indian Idol* and the emergence of mobile publics," in M. Kraidy and K. Sender (eds) *Real Worlds: The Global Politics of Reality Television*. London and New York: Routledge.

Rivero, Y. (2009) "Havana as a 1940s–1950s Latin American media capital," *Critical Studies in Media Communication*, 26(3): 275–293.

Sergie, M. A. and Critchlow, A. (2010) "Saudi Alwaleed, Murdoch discuss alliances," *Wall Street Journal*, 16 January.

Snobar, A. (2008) *Arab Reality TV shows*. Amman, Jordan: Arab Advisors Group.

Straubhaar, J. (2007) *World Television: From Global to Local*. Los Angeles and London: Sage Publications.

Taylor, C. (1999) "Two theories of modernity," *Public Culture*, 11(9): 153–174.

Chapter 3

Watching TV in Havana

Revisiting the Local/Global Television Past through the Lens of the Television Present

Yeidy M. Rivero

When I traveled to Havana, Cuba, I had a very specific evening television routine. I began with the one and a half hour show, *Mesa redonda*, a politically dogmatic program that, in addition to addressing national and regional issues, devotes significant attention to *los imperialistas Yankees* (Yankee imperialists). I then watched locally produced shows including the evening news, a *telenovela*, and culture-oriented programming. On some week nights when I was able to stay awake until midnight or so, I enjoyed subtitled episodes of *Grey's Anatomy*, *The Sopranos*, or whatever Hollywood series was being broadcast. Occasionally, during the afternoons, when my allergies forced me to take a break from old newspapers and magazines, I joined my compañeras/os (comrades) at the institute where I conducted research and watched whatever Brazilian *telenovela* was on.

Locally produced programming in contemporary Cuban television, as it has been since the early 1960s, promotes the benefits of the Cuban revolution and of socialism. Politically oriented panel shows such as *Mesa redonda*, massively attended rallies, educational programs, and culturally oriented shows dominate the television schedule. Conversely, foreign-made television products are more reminiscent of the global dynamics that defined 1950s Cuban and Latin American commercial television.

The interplay between local and global programming in contemporary Cuban television tells a story. This story goes beyond the *telenovelas* and US movies' diegesis, or *Mesa redonda*'s attacks on the US. Contemporary Cuban television is a palimpsest, a technological-cultural artifact that circuitously interconnects Cuba's political, economic, and cultural past and present. "The palimpsest," as Quiroga (2005, pix) writes, "does not reproduce the original, but it dismantles it, writes on top of it, allows it to be seen." As a palimpsest, Cuban television and its programming disclose the multiple connections that have defined this Caribbean island and its media systems. Shaping the foundation of the television/palimpsest is Cuba, a nation state that during the 20th century and through drastic political and socio-economic transformations has been linked to the West/First World and the East/Second World while remaining in the Third World/developing world periphery. Since the arrival of television in Cuba in 1950, the medium has been a chronicler of the nation state and its local, regional, and global exchanges.

In this chapter I examine the interconnection between the local and the global in Cuba's television past by means of genres and programs that are part of the present. In visiting the past, I pay particular attention to the early years of the revolution, a moment of transition when the commercial culture that characterized the medium since its origins coalesced with the pro-socialist culture that began to define Cuba's broadcasting after the triumph of the Cuban revolution. It is also during this period that the regional and global intersections that originally formed the Cuban mediascape were drastically altered to accommodate a new series of global media connections and representations. Thus, by focusing on a transitional moment in the past, I attempt to call attention to the ways in which "recent layers ... [are] built up over older layers," to echo Straubhaar's (2008, p12) arguments regarding media and cultural practices. Contemporary Cuban television, as a palimpsest, creates layers of meanings that do not dissolve the television/cultural past, but that, to some extent, conceal it. One just needs to "scratch" the surface to see the television pasts hidden in the present.

In analyzing Cuba's television past by "scratching" the surface of some contemporary genres and programs, my purpose here is not to create a genre taxonomy, even as I pay close attention to genres. Instead, using Cuba as a case study, I am calling attention to the ways in which a historical contextualization of television, its genres, and its programs serves as a valuable site of information about local political, social, cultural, and media issues, as well as global geopolitical and media formations. Considering Cuba's unique political history in the region and recent revolutions occurring in Latin America (e.g., the case of Venezuela), a historical look at Cuban television could serve as a starting point for future analysis of how, even in closed societies, television has always functioned as a space that is simultaneously local and global even though, at first glance, it appears to be exclusively local. Locality, as Appadurai (1996, pp178–183) reminds us, is "relational and contextual." As we shall see, in Cuba's television and media cultures, genres and programs operate in relation to the present, but they also bring with them discourses associated with the country's and the television/media's local and global past.

My visit to Cuba's television past begins with the *telenovela*, the most popular genre in Cuba and in the Latin American market. I pay particular attention to a series of articles that were published in the Cuban magazine *En vivo* and that discussed the present state of the Cuban *telenovela*. I then move to the panel show genre, one of the most important types of programs in Cuba since the beginning of television on the island. Here, I analyze the transformation of the Cuban panel show from a place for democratic debate to a location for political indoctrination. The piece ends with an analysis of the predominance of Hollywood in the contemporary Cuban mediascape and the possible implications this trend might have on the ways in which we conceptualize the intersections of the local and the global in the medium's past and present.

The *Telenovela*: The (Im)possibility of Totally Obliterating the Past

In December 2009, the Instituto Cubano de Radio y Televisión (ICRT, Cuban Institute of Radio and Television) began to publish *En vivo*, a magazine that centers on Cuba's broadcasting industry. The first issue was dedicated to the *telenovela*, a genre that, according to surveys conducted by the ICRT's research institute, has been the most popular type of program on Cuban television for more than ten years (Paul 2009). Most of the essays, nonetheless, were not dedicated to analyzing local productions. Instead, the pieces addressed what was described as "the crisis" of the Cuban *telenovela*. Discussions regarding the lack of production quality, deficiencies in the narrative structure, and the absence of trained writers were tackled by some *En vivo* authors as possible causes for the local *telenovela* crisis. In fact, one of the writers questioned the use of the term *telenovela* in Cuba, suggesting that a better label would be "dramatized serials" (Fowler 2009). Yet, regardless of terminology issues, the inaugural *En vivo* publication exposed two main themes. First, Cuban audiences favor *telenovelas,* particularly the Brazilian ones, but also others produced in Latin American countries. Second, the crisis of the Cuban *telenovela* is a product of the television professionals' lack of skills to write and produce this type of program.

Cubans' preference for the Brazilian and foreign-made *telenovelas* can be viewed as an example of what La Pastina and Straubhaar (2005) call "multiple proximities," in which audiences might prefer foreign television products over locally produced ones based on factors that go beyond national culture and identity. In the case of Cuba, matters related to cultural capital (Cubans' high levels of education), production quality (e.g., the Brazilian narrative development and aesthetic qualities), and, more importantly, Cuba's highly politicized and censored television system have probably influenced the audience's selection of the Brazilian and foreign product. As Fowler (2009) observes, "The ongoing loyalty of the Cuban public towards the *telenovela*, and in general, towards the melodrama, confirms a beautiful story of cultural resistance … ." The Cuban case presents, as Fowler (2009) obliquely suggests, an excellent example of the audience's ability to transcend the entire flow of Cuba's propagandistic television/media spectrum and the pro-government messages that are part of Cubans' daily lives—from schools, to street signs, to conversations in public places. Even as Fidel Castro and his revolutionary government have deemed the *telenovela* a "deformation of reality" (as quoted in Nuñez-Machín, 1983, p54), Cubans continue to favor the genre.

While audiences have apparently been able to transcend some of the revolutionary regime's messages, Cuban media professionals present a more complex story. Since the early 1960s, television writers, directors, and producers who remained in Havana after Fidel Castro embraced socialism (1961) were retrained to disregard the capitalistic practices that had defined the first ten years of Cuban television. *Telenovelas*, together with other types of programs of the pre-revolutionary years, were eliminated or transformed by the newly instituted government. To be sure, *telenovelas* were not the first type of program removed from the schedule or drastically altered. For instance, giveaway shows such as *Reina por un día* (the Cuban version of *Queen for a Day*) and programs that focused on the lives of the rich and famous (e.g., *Crónica social*) were taken off the schedule four months after the establishment of the revolutionary government. However, contrary to the aforementioned programs, the *telenovela* had a very specific meaning for Fidel Castro. Among all of the genres and programs borrowed from or influenced by US radio and television, the *telenovela* became synonymous with US imperialism and its dominance in Cuba and across Latin America. As Castro remarked in a 1976 speech:

> In 1959 the radio and television monopoly, with the support of North American companies and national capital, maintained a solid foundation. The selling of scripts and programs to more than ten Latin American countries was already established. Cuba's capitalistic radio [and television] was a source of deformation and ideological penetration, not only for our country, but it extended to other countries as well.
>
> Cited in Nuñez-Machín (1983, p54)

Despite the somewhat predictable Cold War political rhetoric, Castro's words contain elements of truth. To talk about *telenovelas* (and *radionovelas*), as I explain below, is to address the structure of the Cuban commercial broadcasting system, the associations between the US and Cuba, and Cuba's media influence in the region.

The Havana broadcasting influence in Latin America first emerged in the 1940s with the sale of radio scripts throughout the region and the transfer of advertisers, radio directors, and scriptwriters to different countries in Latin America. Much of this exportation of scripts and people, as Fidel Castro implied, was fueled by US corporate investments in the region. Companies such as Procter & Gamble and Colgate-Palmolive created alliances with Havana-based businesses, bolstering Cuba's position as one of Latin America's most important centers for radio scriptwriting (see López 1998). Cuban agencies that affiliated with US industries became part of transnational business and creative nets. Still, not all Havana–Latin American connections were a result of US economic interests. In some instances, Cuban advertisers and media professionals foresaw opportunities in new markets and expanded their business to other countries. In other instances, Latin American advertising agency owners traveled to Cuba in search of advertising and radio professionals.

This movement of Cuban media professionals across Latin America continued into the 1950s television era. The Cuban television industry grew exponentially during its early years and, as a result, its television professionals were able to develop their technical and production skills. Cuban expertise in television, together with the nation state's troubling economy and unstable political climate, led to an exodus of television and advertising professionals before 1959. It was precisely these pre-revolutionary connections and dispersions of the 1940s radio era and the 1950s television era that facilitated the incorporation of Cubans within the Latin American and US Spanish-language media industries during the 1960s. One of the media centers where Cubans had an impact, particularly in the realm of *telenovela* scriptwriting, was Brazil (see Mazziotti 1995, 1996; Sinclair 1999). Globo (the most powerful television network in Brazil) "attracted creative talent, notably Gloria Madagan, who brought to Globo her experience in making *telenovelas* for Colgate-Palmolive in Cuba" (Vink as cited in Sinclair 1999, p67). Thus, bearing in mind the Cuban broadcasting influence across the region during the 1940s and 1950s when, as I argued elsewhere, Havana was a Latin American media capital (Rivero 2009), and the Cuban exiles' impact in the region after the triumph of the Cuban revolution, today's Cuban *telenovela* crisis comes as an ironic twist of (historic) fate.

Progressively, the development of a television system dedicated to using "propaganda to constantly elevate the revolutionary consciousness" (Escalante 1976, p10), together with the massive exodus of experienced television professionals, created a cadre of media practitioners who were either politically unable to produce this type of program or unfamiliar with the genre. Some television writers' lack of knowledge about the *telenovela* and, specifically, about the *telenovela*'s thematic and narrative structure (i.e., the classic romance story) comes from the genre transfigurations that occurred during the 1960s. The *telenovela* was transformed to adaptations of literary and theater classics and originally scripted dramas with pro-socialist themes. The romance story that characterizes the *telenovela* vanished in the 1960s. Even as some Brazilian *telenovelas* began airing in Cuba during the 1980s and while traces of the classic *telenovela* re-emerged in local productions in the mid-1980s (Cue-Sierra 2010), the genre's historical connections with US–Cuba relations' past affected the development of this cultural product in Cuba. The current *telenovela* crisis is then a result of historically situated political processes that go beyond the television industry structure. Cubans, once considered creators, trainers, and leaders of *radionovela* and *telenovela* production, have now fallen to the level of apprentice.

Similar to Fidel Castro's position on the genre summarized above, some communication scholars in contemporary Cuba consider *telenovelas* "escapist"-themed programming that portrays a deformed sense of reality (see Fowler 2009). The division between high and low culture and the politicized environment may (or may not) have influenced these categorical definitions of *telenovelas* as unworthy programming. Nonetheless, one thing is clear. Regardless of the massive socio-political and cultural transformations produced by the Cuban revolution, Cubans enjoy *telenovelas*. Whereas the

ICRT Research Center has administered surveys to understand Cubans' preference for the Brazilian and foreign *telenovela* (Paul 2009), there are no concrete answers for their selections. In other words, even though surveys indicate that production quality and themes might attract audiences, the political culture that characterizes Cuba may have an effect on the interviewees' answers. Whatever the case may be, the locally produced prime-time *telenovela* together with the imported ones broadcast in the afternoon have reintroduced elements of a shared Latin American commercial/popular culture space that, as Martín-Barbero (1995) writes, has defined the region since the early 20th century. Local and foreign *telenovelas* and their so-called "escapist" narrative structure serve as a link that reconnects the Cuban socialist television present with the Cuban and Latin American commercial television past and present.

"*Mesa Redonda*": (Re)formulating the Past

Mesa redonda (*Round Table*) is a panel show broadcast Monday to Sunday, from 6:30 pm to 8:00 pm on Cubavisión, Cubavisión Internacional, Canal Educativo, Radio Rebelde, and Radio Habana Cuba. Similar to its local 1950s predecessors, *Ante la prensa* (*Meet the Press*), *Mesa redonda*, and *Telemundo pregunta* (*Telemundo Asks*), and the US precursor *Meet the Press*, the contemporary *Mesa redonda* involves a host and a series of panelists-experts who discuss important national and regional political, social, economic, and cultural issues. Paralleling the aforementioned programs, *Mesa redonda* is additionally viewed by the local public as a show characterized by its intellectually sophisticated debates. The main difference between the previous Cuban panel shows, the US genre model, and the contemporary *Mesa redonda* is the centrality of a dogmatic political discourse. Beneath the surface of today's *Mesa redonda* lies a genre that has maintained the same format as the US panel show but that, contrary to the US model, has been shaped directly by Cuba's political structure.

The Cuban panel show (as hinted at in the previous paragraph) is an adaptation of the US genre. *Ante la prensa*, an indigenized version of NBC's *Meet the Press*, was the first panel show in Cuba; like the US originator, it became the television place for incisive political commentary. Soon after witnessing the success of CMQ-TV's *Ante la prensa*, other panel shows began to appear on Cuban television. Nonetheless, despite the production of similar types of programs, the 1951 pioneer remained the audience's favorite throughout the 1950s commercial television era.

Ante la prensa and other panel shows' journalistic integrity were promptly tested after Fulgencio Batista's military coup. In May 1952 (just two months after his coup), Fulgencio Batista initiated the censorship of news and public affairs programming, which, through journalists and commentators, criticized the government (Salwen 1994, pp93–107). On several occasions during Batista's presidency, and particularly after 1957 when censorship was mandated by law, the government forced *Ante la prensa* and also the CMQ-TV-produced *Mesa redonda* off the air. These ongoing governmental interventions began to give an aura of credibility to Cuban panel shows. For various television critics and for Cuba's intelligentsia, the Cuban panel show was committed to promoting meaningful interrogation and debate regardless of possible consequences. These political, industrial, and historically situated processes that defined the 1950s Cuban panel show were influential in Fidel Castro's televised political persona.

As early as January 1959, the revolutionary leader relied on television to develop a close connection and ongoing communication with the Cuban people. Combining panel shows and televised speeches, Castro's participation in a single year (1959–1960) ranged from one to eight television appearances per month, with interviews lasting from 30 minutes to 6 hours. In some cases, the program hosts were responsible for extending the television interviews. In most instances, however, Castro requested unlimited on-air time to discuss particular political issues.

More than radio or film, television emerged as the medium of the Cuban revolution. Beginning with the transmission of Fidel Castro and the *barbudos*' (the bearded ones') caravan from Santiago de Cuba until their triumphant entrance into Havana, and continuing with Castro's program appearances and televised speeches, television merged revolutionary imagery with political performance. Sometimes reading, but mostly just elaborating on a particular subject, live television represented an ideal media outlet for Castro's intelligence, wit, improvisational skills, energy, and political persona. Targeting the local–national public, the televised Fidel Castro complemented the photographic and filmic imagery of the Cuban revolution that sprouted on the island and across the globe.

The iconic images of the Cuban revolution—that is, of a bearded Fidel Castro wearing combat fatigues, smoking a cigar, and carrying a rifle next to his *barbudo* rebels who were also wearing fatigues and carrying rifles—became part of the global media stage before 1959. The now famous *New York Times*' 1957 series of articles about the revolutionary army in Sierra Maestra immortalized the image of the rugged combatants. These portrayals continued to dominate the visual-photographic narrations of Cuba after the triumph of the revolution. Globally, magazine covers and newspapers reproduced these representations, creating, as several scholars have observed, a romanticized vision of the revolutionary leaders (Quiroga 2005; see also Guerra 2009). Locally, Cuban photographers treated Fidel Castro and his army as "celebrities," adding what literary scholar Quiroga (2005, p95) refers to as "revolutionary glamour" to the revolution, a glamour that was then appropriated by the state:

> The link between advertising and revolution perversely blurs the distinction between political star treatment and mass rallies. … The realm of the visual transformed the revolution by structuring iconic

> sequences. Events coalesced into a common narrative reflecting productive achievements and the mass concentrations in Cuban political rallies.

Photography and film, Quiroga (2005) writes, became key components of the revolutionary iconography and the state political ideology.

Newsreels and documentaries produced by the Cuban Institute of Cinematographic Art and Industry (ICAIC) also turned into important media sources for the screening of revolutionary imagery. Established on March 24, 1959, the ICAIC represented the first effort of a Castro-led government to develop cultural productions that would combine an ideological, aesthetic, and pedagogical function. Initially producing newsreels and, to a lesser extent, documentaries for local and global audiences, work during the ICAIC's first period (1959–1960) celebrated the triumph of the revolution while also screening the corruption of Fulgencio Batista's regime (Burton 1997; see also Chanan 2004).

In concert with the work of photographers and filmmakers, commercial television produced a multiplicity of images about Fidel Castro and the revolution. On television, however, the oral, kinetic, and "unpredictable" element associated with broadcasting live, the physical co-presence of Fidel Castro and the studio audience, and the temporal simultaneity of transmission and reception provided the audience with the notion of watching the "real" Fidel Castro.[1] Although these representations operated through Castro's constructed political persona (from dress style, to the smoking of cigars, to conversation themes), and were mediated by television's technology (e.g., cameras and microphones), the broadcasts of the revolutionary leader functioned through the code of "liveness" that, as Auslander (2008) observes, characterized 1950s television. In other words, because of the live broadcast, the in-house studio audience (in the case of panel shows) and the television audience watched a spontaneous and real Fidel Castro. These live performances of spontaneity not only made it possible for the revolutionary leader to perform various roles (Castro as the altruistic leader, as the educator, as the common man, as the maximum leader, etc.), but many of the depictions were subsequently captured in photographs. With each televised transmission, audiences witnessed spontaneous aspects of the leader and they could also read and see sequences of Castro's verbal and bodily spontaneity in newspapers and magazines.

The perception of Fidel Castro's spontaneity and realness associated with broadcasting live was also influenced by the conventions and trajectory of the Cuban panel show as a genre. Consider Mittell's (2004, p36) observation regarding the meanings of a genre:

> … in any given moment, a particular set of assumptions may emerge as the foremost incarnation of the genre in cultural circulation. While there will probably be contradictory texts and discontinuous discourses active in the generic cluster, a generic dominant functions as the primary connotative implication of generic terminology.

In Cuba, the culture of censorship associated with the panel shows during the Batista regime promoted an atmosphere of truthfulness and openness to Castro and to all revolutionary leaders who participated in these shows. Through his interactions with the shows' hosts and panelists and via his speeches directed towards the camera, Castro was perceived (based on observations by some journalists and commentators) as communicating on a one-to-one basis (see, for example, Abreu 1959; Nuñez-Rodríguez 1959). By addressing the camera the Castro–camera monologue was rearticulated into a conversation between Castro and the Cuban people. Television and, specifically, panel shows facilitated the perception of Castro's immediacy.

If the panel shows represented Castro as an approachable leader dutifully helping the public understand the transformations enacted by his revolutionary government, the outside-the-studio speeches portrayed him as a larger-than-life figure. Through shots of Castro and of the thousands of people who attended the rallies, the power of the televised speeches resided in the pluralization of the Cuban revolution and in the representation of Castro as a maximum leader capable of moving the masses. In other words, even though Castro was the primary speaker and the center of attention at the televised rallies, visually, the depiction of the various cabinet members accompanying him at the podium and of the multitude of people in attendance represented the revolution as a unified movement guided by numerous leaders and supported by all Cuban citizens. Watched by local and, in many instances, foreign audiences (via foreign correspondence newscasts and the ICAIC newsreels), the televised speeches formed part of the revolutionary imagery designed for local and global viewers.

The combined transmission of panel shows and televised speeches together with the newspapers' and magazines' reproduction of sequences and quotes from Castro's televised appearances positioned Fidel Castro as a ubiquitous leader, a man devoted to the revolution and to the Cuban people. As Pérez-Stable (1999, p9) writes, "Fidel Castro emerged as the bulwark of the revolution—the mediator between the people and the government." Following Pérez-Stable's argument, I would add that in the 1959/early 1960 pre-socialist period, Castro's role as a mediator was created and magnified via television. Through television, Fidel Castro explained the revolution to segments of the Cuban people. In addition, television and live broadcasting functioned as catalysts for the reproduction of Fidel Castro's images and speech citations across a range of media outlets. To use Corner's (2000) theorization of the performance of political figures, the "intensity" of Castro's interaction with television and the theatricality of his political persona made him the center of a televised revolution.

Last, and not least important, the fact that Cuban television was technologically developed and that Cuban producers and technical staff were highly experienced media professionals greatly facilitated Castro's political uses of the medium. As Szulc (1962, p198) observed:

> From the time he took power, Dr. Castro has had a well-developed television industry at his disposal. With six channels in Havana and microwave towers carrying signals to every point of the island, Dr. Castro has been able to bring himself and his revolution into the homes and everyday lives of Cubans everywhere.

While Szulc's assumption about the television signal's reach was somewhat inaccurate (in 1959 there were approximately 350,000 television sets in Cuba, most of which were located in the capital and in economically developed regions), having television's latest technological innovation and a team of highly experienced media professionals greatly facilitated Castro's political uses of the medium. Whether in the studio, across the country, or in the US (e.g., the transmission of Castro's participation on NBC's *Meet the Press* in 1959), technology became a tool at the service of Fidel Castro. The US-style commercial/capitalistic technological, economic, and programming structure that defined Cuban television during the 1950s was pivotal to Fidel Castro's political self-promotion and to his selling of the Cuban revolution. Thus, while *radionovelas* and *telenovelas* were cast out as symbols of US imperialistic practices, the Radio Corporation of America (RCA) television technology that enabled the transmission of Castro's speeches across the island and the borrowed US panel show genre that in one particular case had the same name as a US program was re-conceptualized and accepted as tools and channels for creating a new revolutionary television.

Officially, this renewed television system began in October 1960 after the Ministerio de Recuperación de Bienes Malversados (the Ministry for the Recovery of Ill-Gotten Goods) confiscated CMQ-TV, the last commercial network standing on the island. Whereas initially, the government exclusively targeted corporations and people who had (or allegedly had) connections and/or business arrangements with Fulgencio Batista's government, by the end of 1960 all networks were nationalized regardless of the media owners' political affiliation or economic legitimacy. In less than two years, Cuban television moved from being a commercial enterprise connected to the US and Latin America to being a state-controlled system influenced by the Soviet Union broadcasting model. New station names (e.g., Televisión Revolución), high culture-oriented programming (e.g., theatre and ballet), and, after 1961, cultural events and programs imported from countries of the socialist block (such as China, Poland, and the Soviet Union) began to redefine Cuba's television schedule. At the conception of a new Cuban nation state, television became a pivotal ideological piece. Within this new television formation, panel shows were catapulted to a privileged position in television programming.

First seen as a location for incisive political interrogation that resulted in governmental censorship during Fulgencio Batista's regime, and later reconstructed as a television platform for Fidel Castro's political performance, Cuban panel shows endured dramatic ideological transformations to survive what today, in 2013, amounts to 62 years of television existence. The continuing daily broadcast of the contemporary *Mesa redonda* and Fidel Castro's frequent appearances on the show during his years as the nation's leader (1959–2006) attest to the centrality of the genre in Cuba's television culture. Today, Fidel Castro is indirectly present through his *reflexiones*, a series of articles written by the *comandante*, published in the *Granma* (Cuba's official newspaper), and subsequently read by the host of *Mesa redonda* at the beginning of each program. The reading of Castro's *reflexiones* by *Mesa redonda*'s host is indicative of the symbiosis between the revolutionary leader and the Cuban panel show genre. To examine the Cuban panel show is to uncover the vestige of a long television tradition that moved across time periods, global and local formations, and, thanks to the particularities of its genre's history in Cuba, was also able to move across the ideological political spectrum.

What is Your Favorite US Show? *La Anatomía de Grey or 24?*

After imported and local *telenovelas*, the second most popular programs on Cuban television are Hollywood-made television series and movies. It does not matter that these US products are broadcast late at night or that the Hollywood movies are scheduled during the weekends, many Cubans patiently wait to enjoy this programming. As a Cuban friend once told me, "I wonder what Gillian Anderson [star of *The X-files*] would say if she knew that *la Scully* had a forceful fan-club in Cuba." In fact, a researcher affiliated with the ICRT disclosed to me that, based on a nationwide survey conducted between 2006 and 2007, the most popular show in the Cuban provinces was *Friends*. To be sure, Hollywood films have been broadcast on Cuban television since the late 1970s/early 1980s on TV programs dedicated to cinema appreciation (email communication, anonymous source, August 29, 2010).[2] On these shows, the host, (usually a film critic) introduces the film (focusing on the plot and the director) and provides a reading foregrounding the political undertones of the text (see Pertierra 2009). Nonetheless, what is more recent on Cuban television is the broadcast of Hollywood television series. Even though there are no available sources or research about why the ICRT decided to broadcast Hollywood television shows, participants in Pertierra's ethnographic audience study locate the post 1990s as the period when US television began to appear on Cuban airwaves. In their memories, as Pertierra notes, "there is a before and after" informed by the collapse of the Soviet Union (A. C. Pertierra, email communication, September 1, 2010). Additionally, she observes that:

> It is commonly believed in Cuba that the TV shows being broadcast are being downloaded or recorded from commercial satellite channels in the

> region, and that this is commonly believed to be why the number of Hollywood TV shows has greatly expanded since the 1990s.
>
> A. C. Pertierra (email communication, September 1, 2010)

The availability of new technologies and the apparent "payback pleasure" (as someone once told me) of airing US programs without paying copyright fees have apparently contributed to the (re)appearance of Hollywood television on Cuban television.

The aforementioned information regarding the popularity of US programming or the resources used to obtain Hollywood products is not disclosed in *Mesa redonda*. Nor is the fact that, since the early to mid-2000s, people with illegal cable hookups have been disappearing inside their homes to watch shows such as *Caso Cerrado*, a *Judge Judy*-style program with a Cuban-American host, broadcast on the US Spanish-language network Telemundo. Also not disclosed is the fact that many Cubans stay up to date with the latest Hollywood blockbusters and TV shows by renting these products at underground video rentals operating across the island. What is certainly discussed in the national news are the sporadic police raids to close down illegal video-DVD rentals and to dismantle clandestine cable hookups. Thus, while during the late 1950s (when Hollywood began to dominate the Cuban primetime schedule) journalists addressed the problem of television programs and movie importations, particularly regarding potential threats to Cuban employment, the contemporary state-run media structure has preferred to ignore the ideological contradictions of criticizing US imperialism while concomitantly broadcasting Hollywood products.

Global media flows that have characterized other parts of the world for decades have become part of contemporary Cuba, a nation state that until the early 1990s was a closed society politically and economically connected to the countries of the socialist block. The consumption of local, pro-socialist television shows such as *Mesa redonda* and of foreign TV programs such as *24* presents Cubans with a plurality of political, cultural, and economic worldviews. New or, more accurately, historically reconstituted global formations inform today's post-Soviet Cuba material and popular culture conditions. Pertierra (2009, pp116–117) observes as much, noting that:

> As Cubans lost the socialist connections that had previously informed many of their opportunities for travel, their television and musical choices, and the material culture of their homes, new global configurations have become important, linking Cuban residents to tourists and emigrants living in the United States, Canada and Western Europe.

The dismantling of the Soviet Union and the beginning and development of what is generally referred to as the Special Period (the 1990–2000s economic crisis) are the direct and indirect forces behind these local and global intersections and the reconceptualization of regional media flows.

The collapse of the Soviet Union after 1991 radically destabilized Cuba's economy and the lives of Cuban citizens (see Pérez-Stable 1999; Hernández-Reguant 2009). While, in one way or another, post-revolutionary Cuba had always suffered from a shortage of material goods, by the 1980s the nation's citizens had access to basic living needs and there was a general expectation of a soon-to-be "comfortable standard of living" (Gordy 2006, p392). Thus, the 1990s crisis represented an unexpected turn. Traumatic day-to-day survival experiences exemplify the stories coming from those who lived during the Special Period. Elián Gonzalez and his mother might be the most famous *balsero* (rafter) story in the US and globally, but their story provides only a glimpse of the many more who tried to leave the island and lost their lives in the process.

To improve the economic situation produced by the elimination of trade exchanges with former socialist countries, the Cuban state initiated a series of measures that challenged the principles of the revolution as a capitalistic-free, egalitarian society. Legalization of the dollar and of self-employment in certain professional areas, together with the opening of Cuba's economy to foreign investors, began to create class divisions based on access to dollars (or foreign currency) via family in the US and tourist-related ventures (Gordy 2006). Legalized businesses such as *paladares* (semi-private restaurants) and *casas particulares* (semi-private homes where foreigners rent rooms), together with illegal underground jobs such as the selling of Cuban goods on the streets and prostitution, represent some of the tactics used for economic survival. It is within the realm of an illegal/underground economy that the video/DVD rental businesses and the cable hookups began to operate in Cuba. Still, in addition to the availability of foreign programming on national television and via illegal venues, other local–global forces are shaping the contemporary Cuban mediascape.

The incorporation of the Internet in Cuba—albeit restricted, constantly monitored, and mostly available in the workplace and in hotels—adds another layer to the flow of information and media. Cubans with access to the technology use blogs and websites to express their individual-artistic desires, social views, and political hopes. As Venegas (2010) notes, the Internet has become a space where some citizens envision new ways of imagining themselves and socialism. Nonetheless, while the Internet has provided a venue for the representation of a relatively more open (artistically speaking) yet still politically constrained "self," the Cuban state continues its proactive policing role, employing the technology as a tool for revamping a communal, pro-socialist rhetoric. "At present, the Cuban state uses national campaigns, policies, and the internet to emphasize the extent and dimension of the U.S. embargo" (Venegas 2010, p94).

Cuba's TV/media/digital environment interconnects a local-socialist ideology to a global-capitalistic vision, inserting a new layer, to follow Straubhaar's (2008, p12) geological analogy, to the commercial and public service models that delineate most media systems around the world. The contemporary Cuban television/media palimpsest combines and reinvents the commercial and socialist past of the late 1950s and early 1960s transitional era with the socialist and somewhat denied yet definitively palpable capitalistic/commercial present of the post-Soviet Union era. The reality of a political, economic, and cultural transition shapes these two historically distinct moments.

Today, the particularities of Cuban television and media cultures complicate our understanding of globalization, media flows, and consumption practices. Whereas theories of cultural and multiple proximities could answer some questions about the popularity of foreign media artifacts in Cuba, other theories that address the proliferation of global media flows and contra-flows (Thussu, 2007), or the process of cultural appropriation and indigenization (Buonanno, 2008), are not enough to help us understand the cultural, political, and media implications of having a socialist and capitalistic programming/media spectrum in a country that embraces socialism. Nor can these theories help us to comprehend the ways in which the years of interconnections between Cuba and the Soviet Union—the big elephant in the Latin American room—add another layer to the TV/media spectrum of the regional and global past and present. Thus, for example, considering that, according to Thussu (2007, p3), the global "in essence means Western, or more accurately American," how did the media presence of the Soviet Union and countries from the socialist block in Cuba alter our conceptualization of globalization? How did the Soviet Union create another kind of media empire with its own sets of flows? Furthermore, given that, as I discussed in previous pages, the inclusion or rejection of television genres in socialist Cuba did not necessarily correspond to issues of genre origins or content alone, what factors made a genre "pro-communism" in Cuba and how does this categorization compare with, for example, Poland? How are genres and programming practices being altered in Cuba (and other countries of the socialist block) since the collapse of the Soviet Union? How can we explain the ongoing fascination with Hollywood? Needless to say, in Cuba, cultural imperialism *is not the answer*. The Cuban case challenges us to devise new ways to analyze the mediation of the global by considering not only the formations of the present but also how the sedimentations of the past shape this present.

Notes

Thanks to Anna Cristina Pertierra and Cristina Venegas, who provided valuable information about contemporary television and media practices in Cuba.

1 My examination of Fidel Castro's live performance has been influenced by Auslander's (2008) analysis of live and mediatized performance and what he deems as the similarities between early TV and theater.

2 The temporally specific inclusion of US products mentioned by my anonymous source coincides with Varis's (1985, p27) findings, which indicate that in 1983, 23 percent of foreign programming in Cuba came from the Soviet Union, followed by the US with 22 percent. In contrast, ten years earlier, there were no US importations on Cuban television.

References

Abreu, O. (1959) "Fidel Castro en 'Comentarios económicos'," *Cinema*. Anuario, 4 October, CMQ-TV. Instituto Cubano de Radio y Televisión (ICRT), Havana, Cuba.

Appadurai, A. (1996) *Modernity at Large: Cultural Dimensions of Globalization*. Minneapolis: University of Minnesota Press.

Auslander, P. (2008) *Liveness: Performance in Mediatized Culture*. New York, NY: Routledge.

Buonanno, M. (2008) *The Age of Television: Experiences and Theories*. Bristol, UK: Intellect Ltd.

Burton, J. (1997) "Film and revolution in Cuba: The first twenty five years," in M. T. Martin (ed) *New Latin American Cinema*, Volume 2. Detroit, MI: Wayne State University Press, 123–142.

Chanan, M. (2004) *Cuban Cinema*. Minneapolis: University of Minnesota Press.

Corner, J. (2000) "Mediated persona and political culture," *European Journal of Cultural Studies*, 3(3): 386–402.

Cue-Sierra, M. (2010) "Panorámica histórica de la telenovela en Cuba," *Portal de la Televisión Cubana*, 7 August, http://www.tvcubana.icrt.cu/noticias/panoramica-historica-de-la-telenovela-en-cuba/.

Escalante, C. (1976) *Sobre la propaganda revolucionaria*. Havana: Departamento de Orientación Revolucionaria del Comité Central del Partido Comunista.

Fowler, V. (2009) "¿Qué hacer con la telenovela cubana?" *En vivo*, http://www.envivo.icrt.cu.

Gordy, K. (2006) "Sales + economy + efficiency = revolution: Dollarization, consumer capitalism, and popular responses in special period Cuba," *Public Culture*, 18(2): 383–412.

Guerra, L. (2009) "'To condemn the revolution is to condemn Christ': Radicalization, moral redemption, and the sacrifice of civil society in Cuba 1960," *Hispanic American Historical Review*, 89(1): 73–109.

Hernández-Reguant, A. (ed) (2009) *Cuba in the Special Period: Culture and Ideology in the 1990s*. New York, NY: Palgrave Macmillan.

La Pastina, A. C. and Straubhaar, J. (2005) "Multiple proximities between television genres and audiences," *Gazette: The International Journal for Communication Studies*, 67(3): 271–288.

López, O. L. (1998) *La radio en Cuba*. Havana: Editorial Letras Cubanas.

Martín-Barbero, J. (1995) "Memory and form in the Latin American soap opera," in R. Allen (ed) *To Be Continued: Soap Operas around the World*. London: Routledge, 276–284.

Mazziotti, N. (ed) (1995) *El espectáculo de la pasión: Las telenovelas latinoamericanas*. Buenos Aires: Ediciones Colihue.

Mazziotti, N. (1996) *La industria de la telenovela: La producción de ficción en América Latina*. Buenos Aires: Paidós.

Mittell, J. (2004) *Genre and Television: From Cop Shows to Cartoons in American Culture*. London: Routledge.

Nuñez-Machín, A. (ed) (1983) *Pensamiento revolutionario y medios de difusión masiva*. Havana: Editora Política.

Nuñez-Rodríguez, E. (1959) Verdades. *Carteles*, 40, 30 August.

Paul, S. (2009) "Telenovelas extranjeras en la televisión cubana," *En vivo*, http://www.envivo.icrt.cu.

Pérez-Stable, M. (1999) *The Cuban Revolution: Origins, Course, and Legacy*. New York, NY: Oxford University Press.

Pertierra, A. C. (2009) "Private pleasures: Watching videos in post-Soviet Cuba," *International Journal of Cultural Studies*, 12(2): 113–130.

Quiroga, J. (2005) *Cuban Palimpsests*. Minneapolis, MN: University of Minnesota Press.

Rivero, Y. M. (2009) "Havana as a 1940s–1950s Latin American media capital," *Critical Studies in Media Communication*, 26(3): 275–293.

Salwen, M. B. (1994) *Radio and Television in Cuba: The Pre-Castro Years*. Ames, IA: Iowa State University Press.

Sinclair, J. (1999) *Latin American Television: A Global View*. Oxford: Oxford University Press.

Straubhaar, J. (2008) "Global, hybrid or multiple? Cultural identities in the age of satellite TV and the internet," *Nordicom Review*, 2, http://www.nordicom.gu.se/common/publ_pdf/270_straubhaar.pdf.

Szulc, T. (1962) "Cuban television's one-man show," in R. L. Shayon (ed) *The Eighth Art*. New York, NY: Holt, Rinehart & Winston, Inc., 197–206.

Thussu, D. K. (2007) "Mapping global media flow and contra-flow," in D. K.Thussu (ed) *Media on the Move: Global Flow and Contra-Flow*. London: Routledge, 10–29.

Varis, T. (1985) *International Flow of Television Programmes*. Paris: UNESCO.

Venegas, C. (2010) *Digital Dilemmas: The State, the Individual, and Digital Media in Cuba*. New Brunswick, NJ: Rutgers University Press.

Chapter 4

After Bollywood

Diasporic Media in an Age of Global Media Capitals

Aswin Punathambekar

Declaring that 2007 to 2008 had been the year of South Asian entertainment, Neal Shenoy, founding member of the South Asians in Media and Marketing Association (SAMMA), kicked off the 2008 SAMMA Summit (October 31 to November 1) held at the Time Warner Conference Center in New York City. In a glitzy room packed with over 300 industry professionals from film, television, advertising, and digital media companies, Shenoy set the stage for a day-long series of panel discussions focused on the growing influence of South Asian media and the increasingly prominent role that South Asian and South Asian-American media professionals were playing in shaping media and entertainment on a global scale. As images of Padma Lakshmi (*Top Chef*), Sendhil Ramamurthy (*Heroes*), Mindy Kaling (*The Office*), Kal Penn (*The Namesake*), and other instances of South Asian representation in American film and television flashed behind him on a large screen, Shenoy delivered his opening lines:

> We learned that South Asians could play heroes outside of Bollywood films, that we could evolve to far more substantive and accurate depictions of our culture beyond the irate taxi driver and the equally irate terrorist, that good books with universal storylines that feature India as a character could make exceptional cinema. This year, we learned that Sylvester Stallone, Arnold Schwarzenegger, and "I am Legend's" [sic] Will Smith all declared that I am, in fact, Bollywood. Steven Spielberg recognized that Reliance Entertainment's capital actually comes with a potential consumer base of 1.2 billion new ET fans. And we read in the now prophetic words of *Fortune* magazine that one should short Facebook and buy Bollywood.

Bollywood did, in fact, dominate proceedings for the rest of the day. While one panel (Bollywood meets Hollywood) brought together a Mumbai-based filmmaker and executives from Disney, Sahara One, and Gotham Entertainment to discuss relations between Indian and American media corporations, a second panel (It's Not Your Dad's Bollywood) focused attention on diasporic entrepreneurs who had established digital media companies with the goal of creating new trajectories and models of circulation for Bollywood films and film music outside India. In addition, convention organizers had roped in Mumbai-based Pradeep Guha, former CEO of Zee Telefilms, to deliver a keynote address that would provide an overview of ongoing changes in the Mumbai-based film and television industries.

From the perspective of the organizers and attendees, this focus on the media industries in Mumbai was not surprising at all. Indeed, it is no mere coincidence that an organization such as SAMMA has been created at a time when Bollywood is making its presence felt in American public culture and Hollywood is reckoning with Mumbai's position as an influential node of media production and circulation in the world today. For instance, the widely reported entry of Mumbai-based Reliance Entertainment into Hollywood territory through an acquisition of a 250-theatre exhibition chain and investment of US$600 million in Hollywood studio DreamWorks is part of a broader shift in Bollywood production companies' efforts to tap into and leverage new circuits of capital (Siegel and Thompson 2008). Building on shifts in state policy and broader political and economic changes, other prominent Mumbai-based production companies (e.g., UTV and Yashraj Films) have also invested considerable effort in "corporatizing" overseas distribution and exhibition in order to institutionalize an increasingly lucrative overseas box office, particularly in the US and UK. What implications do these changing conditions of media production and circulation hold for the domain of diasporic media? What happens to diasporic media when Indian film and television corporations begin to define and dominate the diasporic mediascape? How are diasporic media entrepreneurs responding to and shaping links between Mumbai and Los Angeles?

This chapter tackles these questions by focusing attention on two recent diasporic media initiatives: MTV-Desi, a television channel that sought to target South Asian-American youth but only lasted 18 months, and Saavn.com, a New York-based digital media company that has emerged as one of the most prominent distributors of Bollywood programming outside India. To map the complex connections across multiple spatial scales that define the imaginations and practices of media professionals at these two ventures, I draw on anthropologist Anna Tsing's reflections on *scale* and *scale-making*. For Tsing (2005, p58), scale is the "spatial dimensionality necessary for a particular kind of view, whether up close or from a distance, microscopic or planetary." A scale, furthermore, is to be understood not just as a "neutral frame for viewing the world," but rather as something that "must be brought into being—proposed, practiced, evaded, as well as taken for granted" (Tsing 2005, p58). Examining MTV-Desi and Saavn.com as scale-making projects, I ask: What does it take to conjure the diaspora as a viable scale of media production and circulation in an age of global media capitals?

An emphasis on questions of scale allows us to acknowledge the growing influence of media capitals that function, as Michael Curtin (2007, p111) observes, "less as centers of national media than as central nodes in the transnational flow of culture, talent, and resources." At the same time, it also highlights the need to account for the continued relevance of the national as a key scale of media production and circulation that structures diasporic media cultures in important ways (Sassen 2006; Govil 2009). As we will see, Tsing's approach opens up new ways of understanding how media industries reimagine their geographic reach. The goal, then, is to examine MTV-Desi and Saavn.com as a way of understanding both the possibilities and challenges facing diasporic media entrepreneurs as they negotiate a decidedly new phase of ties between two major media capitals. I begin, therefore, by drawing attention to the ways in which globalization of the media industries in India, and broader socio-political and economic transitions over the past two decades, have transformed conditions for diasporic media production.

Between Mumbai and LA: Diasporic Media and Desi Culture

> Those who wish to stay, those who have moved, those who wish to return, and those who choose to stay rarely formulate their plans outside the sphere of radio and television, cassettes and videos, newsprint and the telephone. For migrants, both the politics of adaptation to new environments and the stimulus to return are deeply affected by a mass-mediated imaginary that frequently transcends national space.
>
> Appadurai (1996, p6)

Over the past two decades, media scholars have built on the work of social theorists such as Appadurai (1990) and Hall (1990) to provide us with a rich vocabulary and set of tools to analyze the relationship between media and migration. And "diaspora" and diasporic media production have been privileged sites for understanding the shifting, often disjunctive, relations between cultural production, geography, and identity. In the South Asian context, it is possible to now trace an arc beginning with Marie Gillespie's (1995) analysis of media use in a predominantly Punjabi community in a London neighborhood, through Sunaina Maira's (2002) exploration of Indian-American youth culture in New York City, to Shalini Shankar's (2008) ethnography of Desi youth culture during the tech boom in Silicon Valley as a way to foreground transformations in understandings of South Asian diasporic identity and the South Asian mediascape. Desi, which means "from the homeland," is a term that is increasingly used to refer to people of South Asian origin in North America and, as Shankar (2008, p4) points out, "signals the shift from South Asians as immigrants longing to return to a homeland to public consumers and producers of distinctive, widely circulating cultural and linguistic forms."

Where Gillespie documented the creative and strategic ways in which youth in diasporic communities drew on media and popular culture to initiate dialogues between their parents' ideas of culture and their experiences in British society, Maira illustrated how a range of media (from mainstream Indian films to subcultural remix music) functioned as triggers to discussions and contests over broader issues of ethnic authenticity and cultural hybridity, assimilation and race relations, multiculturalism and citizenship. While these issues remain deeply relevant in the lives of Desi teenagers whom Shankar documents and analyzes, her ethnography also illustrates the extent to which being and becoming Desi has changed as South Asians from diverse class, linguistic, religious, and geographic backgrounds have established themselves in places such as California. Arguing that it is no longer productive to characterize second-generation youth as being "culturally and intergenerationally conflicted"—of being "American" at school, "Indian" at home, and "caught in limbo" between these two worlds—Shankar (2008, p6) asserts that Desi youth now "exhibit a far more nuanced consciousness about what it means to be Desi." As we will see, the heterogeneity of Desi youth culture that scholars like Shankar foreground poses a formidable challenge to media executives' efforts to forge a "Desi youth" demographic.

Furthermore, where the South Asian diaspora is concerned, popular and scholarly accounts have tended to privilege cinema, Hindi-language films, and English-language diasporic films, in particular, over print, television, and other forms of cultural production. While one could argue that this provides too narrow a template for understanding the relationship between media and diasporic identity, there are at least three reasons for the privileged position that cinema occupies. The first is simply the enduring popularity of films and film music among South Asian families. A second reason that films and film music figure prominently in discussions of Desi youth culture relates to Desi youth appropriating and remixing film songs and dance sequences in college events, dance clubs, and so on (Maira 2002). Third, it is in and through cinema that diasporic writers and directors such as Hanif Kureishi, Mira Nair, and Gurinder Chadha began addressing the complexities of claiming and defining South Asian identities in countries such as the UK and the US (Desai 2004). However, in these analyses, relations between diasporic media entrepreneurs living and working in cities such as London or New York and media capitals such as Mumbai remain largely unexamined. Why is this so?

Consider, if only briefly for now, diasporic filmmakers such as Mira Nair and Gurinder Chadha. What importance did Bombay, as a center of media production, have for Nair during the 1980s and early 1990s when she produced films such as *India Cabaret* (1985), *Salaam Bombay* (1988), and *Mississippi Masala* (1991)? What did Bombay, as the site of a creative industry whose output she drew on and rearticulated in films such as *Bhaji on the Beach* (1993), mean for UK-based Gurinder Chadha? To be sure, Bollywood has shaped diasporic filmmaking in multiple ways. As Desai (2004, p42) writes:

> The impact of Bollywood on South Asian diasporic filmmaking is multifold. One primary example is the frequency with which Bollywood is

> referred thematically within the films themselves ... in addition, Bollywood conventions are reflected in the aesthetic forms and narrative structures in a variety of films ... also, there is crossover in terms of performers: Shashi Kapoor, Zohra Sehgal, Om Puri, and Shabana Azmi are all actors who have appeared in Indian and diasporic productions. Finally, diasporic filmmakers have employed the networks of distribution that circulate Indian films.

Diasporic media production has always occupied a space between "national" media capitals, between Mumbai and Los Angeles in the South Asian context. However, I would argue that for diasporic media producers, pre-liberalization Mumbai had no specificity either as a "switching point" for capital or as a center of media production capable and/or interested in mediating the experience of migration and diaspora (Curtin 2003). It is this dynamic that changed in significant ways as India embarked on a program of economic liberalization during the late 1980s and early 1990s. For our purposes here, there are three inter-related shifts and emergences to note which defined this socio-historical conjuncture: the growing cultural, political, and economic influence of the Indian diaspora on different spheres of life in India, and conversely the growing influence of Indian media in the diaspora; the transformation of film, television, and advertising industries in cities such as Mumbai, Chennai, and Hyderabad with the entry and establishment of transnational media corporations; and the state's creative responses and efforts to refigure its relationship with both the Indian diaspora and the media industries. While a detailed analysis of these transitions is not possible here, it would be useful to outline how these shifts have reconfigured the field for diasporic media production and circulation.

To begin with, several scholars have argued that Bollywood films played a major role in mediating the newfound centrality of the diaspora, particularly the late-modern, "first-world" diaspora, to India's navigation of a global economy (Prasad 2003; Virdi 2003). Films such as *Dilwale Dulhania Le Jayenge* (*DDLJ*, 1995, Aditya Chopra), *Pardes* (1997, Subhash Ghai), and *Kabhi Khushi Kabhie Gham* (*K3G*, 2001, Karan Johar), which resonated strongly with viewers in India and abroad and count among the most successful films of the past decade, began exploring the cultural space of Non-Resident Indians in new ways. By exploring and cautiously legitimizing the cultural space of Indian life in the diaspora, films such as *K3G* rendered the diaspora's version of Indianness less transgressive or impure and more as an acceptable variant, and, in doing so, set the stage for the state to re-territorialize Non-Resident Indians and remap the sociocultural boundaries of the "national family." As Sujata Moorti (2005, p49) has shown, such narrative strategies whereby "India and Indians ... imaginatively accommodated the diaspora, encompassing it within the folds of a globalized Indian identity," can also be traced in popular magazines and primetime television programs. Further, these representational shifts need to be seen alongside significant institutional and industrial changes. Where cinema is concerned, the rearticulation of state–cinema ties—specifically, the government decision to grant industry status to the business of filmmaking in 1998—and the complex, contested, and ongoing transformation of the "national" Bombay film industry into Bollywood, a "corporatized" cultural industry intent on re-imagining its geographic reach, has been a major development. And when seen alongside the emergence of influential television companies such as ZEE, Sony, STAR, and Sahara One that have also established themselves in diasporic markets over the past decade (McMillin 2010), it becomes clear that diasporic media producers have to negotiate and grapple with the challenges and opportunities that the growing influence of Mumbai-based media corporations present.

I have framed this discussion by positioning diasporic media and culture between Mumbai and LA, but it is crucial to also take into account the influence that Chennai and Hyderabad, two other centers of media production, wield. Among the most dynamic cities in south India, Chennai and Hyderabad are centers of Tamil and Telugu language film production and home to powerful media conglomerates, including the SUN TV network (Chennai) and the Ramoji Group (Hyderabad). In contrast to Bollywood during the 1990s, Tamil and Telugu cinema did not address diasporic communities or wrestle with the issue of re-territorializing diasporic Indians.[1] Further, the distribution of Tamil and Telugu films across the world remains largely unorganized and defined by informal networks involving merchants and grocery stores that cater to South Asian communities, pirate networks that ensure the availability of DVDs within a few days of the film's release in India, and a large number of streaming video and bit Torrent websites. Television, however, is a different story as McMillin's (2010) overview of the SUN TV network makes clear.[2] Managed by Kalanidhi Maran, member of the powerful Karunanidhi family in Tamilnadu, SUN TV was launched as a Tamil-language channel in 1993. Beginning with just three hours of programming, SUN TV developed into a 24-hour channel by 1995 and went on to expand its line up by adding a news channel in 2000 (Sun News), a music channel in 2002 (Sun Music), and a film-based channel (KTV) in 2004. During this time, the network also expanded into Telugu (Gemini TV), Kannada (Udaya TV), and Malayalam-language (Surya TV) programming. By 2002, audiences in the US had access to SUN TV via the Dish Network, and on other carriers in Australia, New Zealand, the UK, Singapore, Malaysia, China, and Russia. What McMillin's (2010) account of the landscape of Indian television companies' global reach makes clear is the segmented nature of the diasporic audiences and, more importantly, the difficulty of imagining a pan-Desi audience demographic. It is in relation to these transformations in the media industries in India, particularly their growing capacity to determine media flows in the diaspora, that we need to examine initiatives such as MTV-Desi and Saavn.com.

MTV-Desi and the Limits of Diasporic Television

In July 2005, MTV Networks announced the launch of MTV-Desi, a niche channel for South Asian-American youth. Launched with great fanfare, MTV-Desi sought to respond to ongoing changes in South Asian-American culture and to create a space within mainstream media that would speak to the particular experiences of Desi youth. In addition to Bollywood song sequences and Indi-pop music videos, the channel would feature US- and UK-based artists such as DJ Rekha, MIA, and Jay Sean, alongside popular American stars in order to create a "multi-ethnic, multi-genre playlist" that would resonate with Desi youth (Dev 2006). While music would remain the primary focus of programming, MTV-Desi would also develop new segments covering a range of topics related to life in South Asia and the South Asian diaspora worldwide, including original shows such as *Live From*, which would track Desi youth culture in cities across North America and the UK, and hit shows such as *Roadies* from MTV-India. Recognizing the transnational nature of Desi youth culture, writers, producers, and VJs worked hard to define MTV-Desi as a unique site of cultural production that neither mainstream American television nor the India-centric programming on Dish and DirecTV could match.

Declaring that MTV-Desi would soon become the "pop culture destination for desis," Nusrat Durrani, general manager and senior vice president of MTV World, explained: "But more than the music, it is also about articulating the stories from this community—young South Asian-Americans who have grown up in the country, but have not seen themselves on TV" (cited in Mozumder 2005). At the same time, Durrani aimed to fashion MTV-Desi not simply as a channel for South Asian-American youth, but as a space that would showcase South Asian cultural production and invite participation from as diverse an audience as possible. MTV-Desi, furthermore, was part of a larger MTV World initiative that involved channels targeting Korean (MTV-K) and Chinese-American youth (MTV-Chi) (Sontag 2005). As the very first mainstream media initiative that targeted diasporic youth culture, these "hyphenated-identity" MTV channels attracted a great deal of positive attention despite the fact that they were available only through an international programming package on DirecTV's satellite television service. MTV-Desi was part of the "Hindi Direct" package that included five other Indian television channels and cost US$29.99 per month.

Eighteen months later, MTV Networks pulled the plug on MTV-Desi, MTV-K, and MTV-Chi, stating that the premium distribution model failed to attract audiences and, hence, advertising revenues. In press releases and interviews, MTV executives also pointed out that the decision was shaped by a larger process of corporate restructuring that Viacom undertook at the time. On the one hand, the cancellation of the MTV World initiative did not come as a major surprise. As one prominent journalist remarked on the widely read blog of the South Asian Journalists Association, "we published next to nothing on the channel, because I couldn't find anyone who watched the satellite channel: no college students, no twenty-somethings with spare change. And it wasn't just me. All the tastemakers I interviewed—DJs, other music types—said they didn't know any MTV Desi subscribers either" (Venugopal 2007). On the other hand, given the fact that all other attempts to carve out a space for Asian-American programming on television—AZN, American Desi, South Asia World, Pan Desi, and ImaginAsian—had failed or struggled to remain viable, the dismay among Desis and other Asian-American groups was understandable. Relying on advertising and marketing discourse that had, over the past decade, constructed the "Asian consumer" and the Asian-American community as an increasingly important audience demographic, protest letters and petitions suggested that these failures reflected a lack of commitment on the part of mainstream media corporations to develop and sustain Asian-American programming. Although letters to MTV Networks urging the company to keep MTV World alive and to make these channels more widely available did not have any effect, it is possible that they did influence MTV's decision to rethink their content production and distribution model. In December 2008, MTV Networks announced the launch of mtviggy.com, a website for Desi, Chinese, Korean, and Japanese youth across the world. While media attention has moved on, focusing in recent months on American television networks' attempts to create South Asian-themed programming (Andreeva 2010), analyzing this moment of "failure" is crucial for it brings into sharp relief the challenges facing media professionals in imagining diasporic audiences. In this case, it also encourages us to reflect on the limits of television as a site for the articulation of contemporary Desi youth culture.

To begin with, the narrowcasting logics of American television and industry lore concerning minority audiences do provide one explanation as to why MTV-Desi did not succeed. Nusrat Durrani and others at MTV understood very well that the relationship between "diaspora" and "home" was much more ambivalent for Desi youth compared to their parents' generation, and that MTV-Desi could not succeed by mimicking MTV-India or other Indian television channels. However, despite Durrani and others' efforts to position MTV-Desi as a uniquely diasporic space, they were forced to strike a distribution deal with DirecTV and locate MTV-Desi within an India-centric programming package. On the one hand, Durrani and others at MTV Networks were grappling with a changing distribution landscape in the US, uncertain about a channel such as MTV-Desi reaching audiences via satellite television. As Durrani (pers. comm., October 28, 2008) explained:

> When the project was green lit, the distribution landscape was kind of weird. We were still living in the world of linear TV and traditional distribution partners. Cable or satellite companies. And, you know, there was pressure from our linear TV partners, so we partnered with them and launched the channels as premium TV services, and, you know, everybody loved the idea. They said, let's align ourselves to the model

> that exists, you know, the premium TV model, where there is enough hunger within these audiences to actually pay a premium to get the channel.

Durrani and his team were also working in a context in which advertising and marketing professionals had succeeded in constructing Asian-Americans as a consumer demographic that remained untapped. While the intersections of South Asian cultural production and American public culture can be traced through the work of artists such as DJ Rekha and sub-cultures in cities such as New York, Chicago, and Los Angeles, media companies' interest in this hitherto marginalized community was sparked, in part, by the results of the 2000 US census which revealed that median household income for South Asian families was US$64,000. In much the same way that advertisers and marketers worked to commodify Latinos during the mid-late 1990s (Dávila 2001), companies such as the *Evershine Group* took on the task of constructing a Desi demographic. As one widely circulated article entitled "Chasing Desi Dollars" proclaimed (Kiviat 2005):

> There are some 2.5 million desis in the U.S., and the vast majority are Indian. That may not seem terribly significant compared with, say, 40 million Hispanics, but consider how premium a customer a South Asian is: Indians alone commanded $76 billion worth of disposable income last year ... median household income is nearly $64,000—50% higher than the national average.

It hardly needs to be pointed out that this logic of enumeration papers over the diversity of South Asian histories and cultural practices in North America, particularly in relation to second-generation diasporic youth. Even though Durrani and others at MTV-Desi recognized that it would be a mistake to imagine Desi youth's engagement without taking into account linguistic, regional, and other differences that would have shaped their experiences, statements from others at MTV Networks revealed that this was how Desi and, more broadly, Asian-American identity continued to be mobilized. Statements from other industry executives also revealed how this advertising/marketing discourse positioned Desi youth outside the boundaries of American national culture. A particularly telling press release from MTV described the MTV World initiative as an attempt to "tap into the rich transcultural nature of the target audiences in a manner that uniquely connects local audiences to *their homeland*" (Sheth 2005, my emphasis). For Durrani and his team, the struggle to explain and get others at MTV to understand the particularities of the Desi experience extended into other aspects of media production as well. Consider, for instance, their experience in dealing with graphic design firms in New York City:

> So the packaging, created by graphic designers, people working at creative agencies and so on—no one had done work for a South Asian

> channel. They might have approached it from a very stereotypical kind of way, where they again had dancing elephants, sitar music, and incense and we were just not going to allow that. We wanted a very high quality, very global, and edgy look and feel. Just to get that part out of the way—to create something meaningful—was not easy. To create packaging that merged the two identities, American identity and your South Asian identity was not easy, because, not only has it not been done before, it also is dynamic. It's changing, it's ever changing, and what South Asianness meant in this country in the '70s is very different from what it means now.

However, MTV-Desi's failure cannot be explained by focusing on the logics of American television alone. As detailed above, the growing influence of Indian film and television companies in defining media circulation in the diaspora also played a crucial role in shaping the channel's programming, distribution, and reception. It would not be an exaggeration to state that television in India has undergone major changes over the past 10 to 12 years. As Shanti Kumar (2006a) and several other scholars have documented, the establishment of influential transnational networks such as Star TV and translocal networks such as ZEE, Sun, and Eenadu during the 1990s transformed the ways in which television operated as a cultural institution. What began with local cable operators stringing cables across rooftops to connect homes to the new and fascinating world of Star Plus, Star Sports, MTV, and BBC News had, by the mid-1990s, grown into a stable satellite and cable industry with rapidly expanding viewership. Over the past decade, these television companies have invested considerable effort in reaching diasporic audiences (particularly in North America and Western Europe). As Dudrah (2002, p169) points out, taking advantage of the mainstream media's neglect of Asian audiences in Britain and Europe, companies such as ZEE TV quickly established themselves as an alternative, offering a "variety of programs that visibly manifest themselves as South Asian for South Asian viewers." This holds true in the US as well, where a range of Indian television channels targeting different linguistic groups are offered through Dish and DirecTV's satellite television services. However, even a cursory look would confirm that India-centric media, non-subtitled television programming in particular, also ends up marginalizing Desi youth who might have neither the linguistic skills nor the level of immersion in the politics and culture of the Indian subcontinent to engage with its soap operas, sitcoms, and reality shows.

I would argue, then, that MTV-Desi is symptomatic of a larger problem confronting Desi television production—of being caught between the nationalist logics of two powerful media industries. To return to Anna Tsing's (2000) terms, we could also understand MTV-Desi as a project that sought to re-imagine the scale of diasporic media, only to be stymied by the difficulties of working in-between the national and transnational scales of

American and Indian media industries. Given the difficulties of creating programming that cuts across and speaks to the diversity of Desi youth culture, it is also worth asking if we can expect television, in its current form and structure, to serve the needs of a decidedly transnational, hybrid, and networked youth culture. Surely Nusrat Durrani and his team at MTV grappled with this question as they reworked the MTV World initiative into mtviggy.com. Does the dot-com space allow for more flexible and productive links across spatial scales? What might the fortunes of a company like Saavn.com tell us about the work of creating a circuit of media circulation that is able to leverage changing relations between "national," "global," and "diasporic" in ways that MTV-Desi could not?

Saavn.com: Diasporic Entrepreneurs and the Promise of Digital Media

The last panel of the SAMMA Summit of 2008—"It's Not Your Dad's Bollywood"—brought together three diasporic entrepreneurs who had launched digital media companies with ties to Bollywood. Given that the slogan signals a generational break, we might begin by asking: for Vin Bhat (Saavn.com), Anjula Acharia-Bath (Desihits.com), and Geetanjali Dhillon (Jaman.com), what was dad's Bollywood? As the panelists recalled, dad's Bollywood was what they grew up with in their homes and communities in the US: weekend screenings at a community hall or in a university auditorium, often arranged by an enterprising South Asian family; film music played at home or in cars on road trips with other South Asian families; dances performed at community events and on India Night shows in multicultural college campuses across North America; and, most crucially, one-hour programs featuring Bollywood songs on the local Public Access Station every Saturday morning. Supported by advertising from South Asian grocery stores, restaurants, and companies such as Western Union, these television shows were also locally produced or, at best, purchased from the New Jersey-based *Asian Variety Show*. It was not, as the three panelists insisted, *their* Bollywood. Leveraging ties in Mumbai and other key media capitals, including Los Angeles, New York, and London, these entrepreneurs argued that they were uniquely positioned to respond to the transcultural dimensions of diasporic culture and, in the process, establish new trajectories of circulation for Bollywood films and film music in ways that were not possible for professionals working primarily in Mumbai. While all three companies are interesting cases to consider, I focus on Saavn.com here because it is focused exclusively on Bollywood, unlike Desihits or Jaman. In what follows, I draw on an in-depth interview with Vin Bhat in addition to his presentation at the SAMMA panel to narrate the emergence of Saavn.com as one of the largest and most influential Bollywood digital media companies.

The story begins in the early 2000s when Vin Bhat decided, after a few years in the world of investment banking in New York City, to try his hand at being an entrepreneur. With three other colleagues, Bhat launched a software company that focused on servicing media clients and designed contextual advertising. As with several such ventures in the post-boom dot-com economy, Bhat and his colleagues sold their company to a venture capital firm and, in a move that would bring them in close contact with the media world in Mumbai, decided to spend a few months in India. "This was 2003–4 and we didn't really have a plan. It was about taking some time off while also getting to know the media industry in a place that was attracting attention across the world," Bhat recalled. Through contacts at major banks in India, Bhat and his colleagues were able to meet a range of media industry professionals across the film, television, and music sectors in Mumbai. According to Bhat, this experience was formative. In particular, what sparked their interest was the recognition that Mumbai-based professionals were struggling to establish a presence in overseas territories. "When we heard from so many people that it was difficult to get a sense of the [overseas] market when there isn't proper reporting and the distribution chain is opaque, it got us thinking," Bhat recounted. "We wanted to go back to the drawing board, to figure out how to solve this problem in the industry."

Back in the US, Bhat and his colleagues began by approaching cable operators. With the help of former advisors, one of whom had been an executive vice president at Fox Broadcasting and another a vice president at Bravo, they were able to initiate conversations with major companies such as Time Warner Cable early in the fall of 2003. As it happened, their visit took place in the context of cable companies across North America exploring the possibilities of video on demand (VOD) services as a way of tapping into the sought-after "Asian American audience." To be sure, cable television in the US has been a vital space for a range of transnational, ethnic, and exilic media. This has, however, been limited to public and leased access television. As Naficy (1993, p144) and others have shown, exilic and diasporic media producers have had to work "at the intersection and in the interstices of culture industries; transnational, national, federal, state, local, private, ethnic, commercial and non-commercial funding agencies." In 2003, Vin Bhat and his colleagues entered into conversations with companies that now aimed to create a space for diasporic audiences within mainstream cable television. "It was a surprise, yes," Bhat continued:

> Time Warner cable was telling us that customers were emailing call centers and calling customer service asking for Bollywood channels. But of course, in 2004, Time Warner could not carry those channels. Unlike Dish and DirecTV, their issue was one of space, of capacity. So they wanted to get individual Bollywood movies that they could put on VOD, which was a new service they had launched. And this is where we came in.
>
> Vin Bhat (pers. comm., November 10, 2008)

Vin Bhat and his team struck a deal which involved them figuring out a way to get licenses from film producers in Mumbai, devise a marketing campaign for Time Warner, and develop a revenue-sharing agreement. In October 2003, Bhat and his colleagues launched BODVOD Networks in partnership with two New York-based companies, [212]Media and Schramm Sports & Entertainment, and began supplying Bollywood films and other South Asian media content to cable operators and creating marketing plans for cable operators to attract South Asian audiences. Over the next few years, BODVOD expanded to other cable carriers (e.g., Comcast and Rogers) and was able to claim a distribution base of 19 million homes across North America. In addition, having negotiated rights for global distribution, Bhat also explained that their objective was to expand to South Africa, Australia, and parts of the Middle East and Southeast Asia.

Securing distribution arrangements with film producers in Mumbai was not, however, a straightforward affair. When Vin Bhat and others at BODVOD approached various production companies and studios with a revenue-sharing proposal, they found themselves unable to persuade anyone that focusing on the cable television market and, subsequently, online and mobile phone platforms would be just as crucial as intervening in theatrical distribution and exhibition practices. The fact that BODVOD was primarily a New York-based company managed by people who did not have deep ties in the film industry did not help matters either. According to Bhat, one production company finally agreed to a revenue-sharing agreement and this, in turn, made it possible for BODVOD to raise investor capital. These funds were used to pay minimum guarantees to the production company, an arrangement that soon attracted several other producers with interests in the overseas market. "Once we had minimum guarantees in place and got prominent companies like UTV and Adlabs to sign up, things got easier," recalled Bhat and went on to explain that BODVOD positioned itself in relation to the rhetoric of "corporatization" that had come to define the transformation of the Bombay film industry into "Bollywood Inc." Industry professionals in Mumbai, for their part, were interested in the increasingly lucrative overseas territories and were more than pleased with the level of transparency in reporting when it came to VOD. With pirate networks remaining robust and the theatrical distribution chain just as opaque and unreliable, BODVOD seemed to offer a way forward.

Having forged relations with the media industries in India and the US, BODVOD moved beyond Bollywood films and the cable business to enter the Internet and mobile phone sectors as well. Adopting the name "Saavn.com," Vin Bhat and his colleagues established a joint venture with Hungama, a Mumbai-based media company that is one of the largest aggregators of content for the mobile phone across Asia. By 2007, Saavn.com had established a distribution network that included films and film music, and more importantly, an audience network that spanned the globe. This audience, Vin Bhat further claimed, was what everyone involved in Saavn.com regarded as the most important aspect of their business. "In ways very similar to the way BET created a voice ... recognition for African Americans twenty years ago, and Telemundo did a few years back, Saavn.com aims to be that voice for the South Asian demographic," declared Bhat. While Durrani and his team at MTV-Desi found themselves unable to forge links between the "national" scale-making projects that Indian and American media corporations were invested in, Bhat and his colleagues at Saavn.com succeeded by positioning themselves as brokers between Indian and American media companies to forge a fragmented, yet lucrative South Asian-American audience. In other words, Saavn.com could be understood as an instance of a project that succeeded because it was able to "manage tentatively productive linkages with other scale-making projects" (Tsing 2000, p58) while MTV-Desi could not.

Rethinking Diasporic Media

Narrating the fortunes of two recent diasporic media initiatives, this chapter has examined how the restructuring of the media industries in Mumbai, and changing relations between Mumbai and LA, have reconfigured the space of South Asian diasporic media production and circulation. It would be useful to recount some of the broad changes that are at play here: new circuits of capital that have transformed Mumbai's media world over the past 10 to 15 years, and set the stage for the ongoing and contested refashioning of the film industry into a "global" cultural industry; the Indian state's efforts to remake its ties both with the media industries—specifically, the articulation of a creative industries policy beginning in the late 1990s—and the first world, late-modern diaspora; changing relations between LA and Mumbai, which involves both American media corporations' efforts to strengthen their presence in the Indian market, and Indian companies' moves into the US; the imaginations and practices of media professionals grappling with the challenges of negotiating new cultural-geographic terrains; and, finally, the labor of fans/audiences/"pirates" who download, remix, share, and enable new trajectories of media circulation.

Perhaps it is worth asking if, in the first instance, the term "diasporic media" is useful even at a purely descriptive level? MTV-Desi and Saavn.com are, without a doubt, mainstream media ventures and are strikingly different when we consider accounts of diasporic film and television production by scholars such as Naficy (1993) and Desai (2004). Far from being exilic or interstitial, these media initiatives cannot be grasped without accounting for the ways in which relations between global media capitals—in this case, Mumbai and Los Angeles—have begun to define media circulation in the diaspora and structure the conditions for diasporic media production. What conceptual and methodological frameworks could we use to come to grips with these socio-political and industrial shifts and our understanding of contemporary diasporic media cultures?

Patterns of media production and circulation that I have described in this chapter certainly cannot be grasped in terms of a global–local dichotomy that has, as several scholars have pointed out, dominated the study of globalization (see Chapter 2 in this volume). Concerning ourselves with political economy and focusing attention on the structural and regulatory foundations upon which this kind of media production rests, while certainly crucial, would not foreground issues of identity and work at the level of the everyday or, for that matter, sites such as the SAMMA Summit where various imaginations of the "national," "global," and "diasporic" are in play. "Thinking nationally," of the ways in which the "national has given the media industries a way to think both globally and locally" (Govil 2009, p133) is certainly critical, but does not necessarily capture the ways in which nodes of media production are now "bound up in a web of relations that exist at the local, regional, and global levels, as well as the national level" (Curtin 2003, p204). These frameworks and approaches, moreover, do not focus on the "diaspora" as a scale that is not only relational but, in fact, unsettles the "national" and the "global" as pre-given and uniformly imagined scales of operation for the media industries.[3]

Thus, I have drawn on a set of terms that anthropologist Anna Tsing (2000) deploys to conduct ethnographies of global connection: *scale* and *scale-making*. Projects such as establishing and operating a digital media company do not just reveal the interpenetration of scales (national, global, regional, diasporic, urban)—that much, I think, is recognizable from the account of Saavn.com that I have provided. Such projects, Tsing (2000, p57) would say, "cannot limit themselves to conjuring at different scales—they must conjure the scales themselves." Focusing on the rhetorics and practices of scale-making allows us to understand MTV-Desi and Saavn.com as projects that sought to re-imagine the diaspora as a commercially viable scale of media production and circulation. And surely this is what Vin Bhat, Nusrat Durrani, and their colleagues were doing—imagining and mobilizing different visions of the South Asian diaspora, from New York City, as Desis, in collaboration with professionals in Mumbai, Los Angeles, and other cities in the world, building on shifting notions of Desi identity and ongoing changes in Indian and American media industries. The two ventures I have described conjured the Desi diaspora in different ways, and encountered different sets of challenges and opportunities. In the process, they also reveal the contingent nature of claims about "national," "global," and "diasporic" that inform the operations of the media industries today.[4] Institutions, nation states, media companies, and the social worlds of media industry professionals—it is the intersections between various scale-making projects that are invoked in these sites, and instances of success and failure that we need to examine in order to grasp what happens to diasporic media in an age of global media capitals.

Notes

1 For a detailed consideration of why Tamil cinema has not focused attention on diasporic life and culture, see Velayutham (2008, pp172–188).

2 Further, Chennai and Hyderabad's influence needs to be understood beyond the issue of capitalizing on diasporic audiences. For an analysis of Hyderabad's Ramoji Film City and its global reach, see Kumar (2006b). As major hubs of information and technology industries, these cities are also key nodes for a range of post-production processes.
3 For a theoretical overview of the concept of "scale," see Marston (2000).
4 In this chapter, I have focused primarily on mainstream, commercial media production. However, I would argue that issues of scale and scale-making are even more crucial when it comes to mapping and analyzing grassroots, independent media projects as well. Consider, for example, how an artist such as Nina Paley works at the intersections of national and diasporic imaginaries in developing *Sita Sings the Blues*, "a musical, animated personal interpretation of the Indian epic the Ramayana," http://www.sitasingstheblues.com.

References

Andreeva, N. (2010) "Indian-themed comedies a new TV trend," *The Hollywood Reporter*, 31 January.

Appadurai, A. (1990) "Disjuncture and difference in the global cultural economy," *Public Culture*, 2(2): 1–24.

Appadurai, A. (1996) *Modernity at Large: Cultural Dimensions of Globalization.* Minneapolis, MN: University of Minnesota Press.

Applebaum, S. (2007) "In the year of the pig, cable is getting fat on Asian VOD," *Cable 360*, 22 January.

Curtin, M. (2003) "Media capital: Towards the study of spatial flows," *International Journal of Cultural Studies*, 6(2): 202–228.

Curtin, M. (2007) *Playing to the World's Biggest Audience: The Globalization of Chinese Film and TV*. Berkeley, CA: University of California Press.

Curtin, M. and H. Shah (eds) (2010) *Reorienting Global Communication: Indian and Chinese Media beyond Borders*. Urbana-Champaign, IL: University of Illinois Press.

Dávila, A. M. (2001) *Latinos, Inc.: The Marketing and Making of a People*. Berkeley, CA: University of California Press.

Desai, J. (2004) *Beyond Bollywood: The Cultural Politics of South Asian Diasporic Film*. New York, NY: Routledge.

Dev, S. (2006). "She has her MTV(Desi)," *Nirali Magazine*, 1 September, http://niralimagazine.com/2006/09/she-has-her-mtv-desi/.

Dudrah, R. K. (2002) "Zee TV-Europe and the construction of a pan-European South Asian identity," *Contemporary South Asia*, 11(2): 163–181.

Gillespie, M. (1995) *Television, Ethnicity, and Cultural Change*. London: Routledge.

Govil, N. (2009) "Thinking nationally: Domicile, distinction, and dysfunction in global media exchange," in J. Holt and A. Perren (eds) *Media Industries: History, Theory, and Method*. Malden, MA: Wiley-Blackwell, 132–143.

Hall, S. (1990) "Cultural identity and diaspora," in J. Rutherford (ed) *Identity: Community, Culture, Difference*. London: Lawrence & Wishart, 223–237.

Kiviat, B. (2005) "Chasing desi dollars," *Time*, 6 July, http://www.time.com/time/magazine/article/0,9171,1079504,00.html.

Kumar, S. (2006a) *Gandhi Meets Primetime: Globalization and Nationalism in Indian Television. Popular Culture and Politics in Asia Pacific*. Urbana, IL: University of Illinois.

Kumar, S. (2006b) "Mapping Tollywood: The cultural geography of 'Ramoji Film City' in Hyderabad," *Quarterly Review of Film and Video*, 23: 129–138.

Maira, S. (2002) *Desis in the House: Indian American Youth Culture in New York City. Asian American History and Culture*. Philadelphia, PA: Temple University Press.

Marston, S. (2000) "The social construction of scale," *Progress in Human Geography*, 24(2): 219–242.

McMillin, D. (2010) "The global face of Indian television," in M. Curtin and H. Shah (eds) *Reorienting Global Communication*. Urbana-Champaign, IL: University of Illinois Press, 118–138.

Moorti, S. (2003) "Desperately seeking an identity: Diasporic cinema and the articulation of transnational kinship," *International Journal of Cultural Studies*, 6(3): 355–376.

Moorti, S. (2005) "Uses of the diaspora," *South Asian Popular Culture*, 3: 49–62.

Mozumder, S. (2005) "Desi television comes alive," *India Abroad*, XXXV(15), 7 January.

Naficy, H. (1993) *The Making of Exile Cultures: Iranian Television in Los Angeles*. Minneapolis, MN: University of Minnesota Press.

Prasad, M. (2003) "This thing called Bollywood," *Seminar*, 525: 17–20.

Sassen, S. (2006) *Territory, Authority, Rights: From Medieval to Global Assemblages*. Princeton, NJ: Princeton University Press.

Shankar, S. (2008) *Desi Land: Teen Culture, Class, and Success in Silicon Valley*. Durham, NC: Duke University Press.

Sheth, J. (2005) "Desi making waves," *ABCD Lady*, July, http://www.abcdlady.com/2005-07/art5.php.

Siegel, T. and A. Thompson (2008) "DreamWorks, Reliance close deal," *Variety*, 19 September, http://www.variety.com/article/VR1117992505.

Sinclair, J. and S. Cunningham (2000) "Go with the flow: Diasporas and the media," *Television & New Media*, 1(1): 11.

Sontag, D. (2005) "I want my hyphenated-identity MTV," *New York Times*, 19 June.

Tsing, A. L. (2000) "The global situation," *Cultural Anthropology*, 15(3): 327–360.

Tsing, A. L. (2005) *Friction: An Ethnography of Global Connection*. Princeton, NJ: Princeton University Press.

Velayutham, S. (eds) (2008) *Tamil Cinema: The Cultural Politics of India's Other Film Industry*. New York, NY: Routledge.

Venugopal, A. (2007) "Obit, MTV-Desi," *SAJAforum* (Blog), 18 February, http://www.sajaforum.org/2007/02/obit_mtv_desi.html.

Virdi, J. (2003) *The Cinematic imagination: Indian Popular Films as Social History*. New Brunswick, NJ: Rutgers University Press.

Chapter 5

Regional Cinemas and Globalization in India

Shanti Kumar

In February 2005, the *National Geographic* featured a special section on the globalization of Indian cinema with the headline "Welcome to Bollywood." The magazine's lead article titled "Lights, Camera, India!" introduced Indian cinema to the novice reader in the following terms: "For sheer pizzazz—and number of pictures produced—India's over-the-top film industry, known as Bollywood, easily out-hollywoods Hollywood" (Mehta 2005). In a similar vein, *Time* magazine in Asia featured the leading Bollywood actress Aishwarya Rai on its cover page for an issue in 2002 with the headline "The New Face of Film." The inside pages featured glossy images from a photo shoot with Rai, and the photos were accompanied by the following text: "Queen of Bollywood: Indian leading lady and former Miss World Aishwarya Rai leads the charge as a hipper, edgier, more professional Bollywood bids for global dominance" (Perry 2003). In 1999, when BBC News conducted an online poll inviting readers to select the star of the millennium, the Bollywood superstar Amitabh Bachchan was voted the greatest star of stage or screen. He was voted to the top of the list by readers ahead of many well-known British and American stars such as Lawrence Olivier, Alec Guinness, Charlie Chaplin, Homer Simpson, Robert De Niro, Cary Grant, Buster Keaton, and Marilyn Monroe, who came in second through ninth on the list, respectively. The tenth spot on the list went to Bollywood leading man Govinda who during the late 1990s was one of the most popular stars in Hindi cinema (BBC News 1999).

As is evident from these media reports, there has been a great buzz about Indian cinema going global in recent years. Due to its hegemonic status as India's "national cinema," Bollywood has come to represent the global face of Indian cinema. But South Indian cinemas (traditionally defined as regional cinemas) have also been rapidly globalizing, particularly since the 1990s when globalization became a mantra in the Indian film industry. South India is a term that collectively refers to the four peninsular states of Andhra Pradesh, Karnataka, Kerala, and Tamil Nadu in the subcontinent. The term "regional cinema" is used to describe films made in the dominant regional language of each Indian state. For instance, the majority of films in Andhra Pradesh are made in Telugu, which is the dominant regional language in the state. Similarly, Kannada language films in Karnataka, Malayalam language films in Kerala, and Tamil language films in Tamil Nadu constitute the regional cinemas in each of these states. Since most states in India are linguistic states, other regional cinemas in India are similarly indicated in relation to the dominant language of the state. Thus, Bengali cinema is defined as the regional cinema of the state of West Bengal, Gujarati cinema is the regional cinema of the state of Gujarat, Oriya cinema is the regional cinema of the state of Orissa, and so on (Thoraval 2001).

In this chapter, I argue that there is a need to distinguish between the recent media hype about Bollywood as a global phenomenon and the historical realities of globalization in Indian cinema, particularly in the context of "regional" cinemas in South India. I propose to re-imagine regional cinemas as "regional worlds" by examining how South Indian cinemas are rapidly going global and creatively linking ethnic, linguistic, historical, and geographical elements of identity to material and symbolic mobility of people, products, places, and cultures around the world. As Arjun Appadurai (1999) argues, "regional worlds" is a useful term to focus on that productive tension between "region" and "world" and to create new ways of studying "areas" that have been traditionally considered marginal in national and transnational contexts.

In the first section of this chapter, I discuss why regional cinemas in India have been marginal to categories such as "world cinema" and "national cinema," which have traditionally—and quite problematically—been used in film studies to understand film industries and cultures beyond Hollywood in the US.[1] In the second section, I discuss some of the historical reasons for the marginalization of South Indian cinemas as "regional cinemas" by interrogating the hegemony of the Hindi-language film industry, or Bollywood, in relation to the cultural politics of nationalism in India. In the third section, I propose that the recent emergence of global film cities in South India, such as Ramoji Film City in Hyderabad and Innovative Film City in Bangaluru (formerly Bangalore), represents an exemplary case study for understanding how regional cinemas are being remapped as regional worlds in India today. In conclusion, I argue for a critical regionalism in cinema studies that highlights the "regional worlds" framework to explore the many ways in which regional cinemas are at once globally dispersed and tied to particular locations.

Regional Cinemas and Regional Worlds

In an attempt to redefine marginalized regions of the world as "regional worlds," Appadurai (1999) calls for a decisive shift away from "trait geographies" to what he calls "process geographies." He contends that traditional thinking about "areas" or "regions" in the humanities and social sciences has for long been dominated by ideals of coherence in geographical, civilizational, or cultural terms. These research imaginations of coherence,

Appadurai argues, are based on some "trait lists" that define a region or an area in terms of a static set of values, languages, practices of everyday life, and so on.

According to Appadurai (1999, p232), the traditional approaches—however sophisticated in their research imagination—are fundamentally flawed because they look at "areas" and "regions" as "relatively immobile aggregates of traits, with more or less durable historical boundaries and with a unity composed of more or less enduring properties." In contrast, Appadurai favors a new architecture for the study of areas and regions based on "process geographies," which foregrounds "significant areas of human organization as precipitates of various kinds of action, interaction and motion—trade, travel, pilgrimage, warfare, proselytization, colonization, exile and the like" (Appadurai 1999, p232). Emphasizing the need for understanding the large regions and areas on our current maps as dynamic processes and not as permanent geographical facts, Appadurai (1999, p232) writes:

> Regions are best viewed as initial contexts for themes that generate variable geographies, rather than as fixed geographies marked by pre-given themes. These themes are equally "real," equally coherent, but are results of our interests and not their causes. The trouble with much of the paradigm of area studies as it now exists is that it has tended to mistake a particular configuration of apparent stabilities for permanent associations between space, territory and cultural organization. These apparent stabilities are themselves largely artifacts of: the specific trait-based idea of "culture" areas; a recent Western cartography of large civilizational land-masses associated with different relationships to "Europe" (itself a complex historical and cultural emergent); and a Cold-War based geography of fear and competition in which the study of world languages and regions in the United States was legislatively configured for security purposes into a reified map of geographical regions.

Drawing on Appadurai's call to shift our research imagination away from trait geographies to process geographies, I contend that the productive tension between "region and world" in the term "regional worlds" is particularly relevant for understanding the place of South Indian regional cinemas in the world. Traditionally, "world cinema" has been used by scholars in film studies and by filmmakers and film aficionados alike to describe the diversity of cinemas found around the globe. Often the term is used to define the cinematic traditions of different countries around the world in contrast to American cinema—and sometimes more specifically to the Hollywood film industry. In an introduction to *The Oxford History of World Cinema*, Geoffrey Nowell-Smith (1996, pxx) describes his struggle as the editor who had to reconcile these two competing meanings of the term in the book. He writes:

> On the one hand the book tells the history of the cinema as a single global phenomenon, spreading rapidly across the world and controlled,

> to a large degree, by a single set of interlocking commercial interests. But it also, on the other hand, tells the history of many different cinemas, growing in different parts of the world and asserting their right to independent existence often in defiance of the forces attempting to exercise control and to "open up" (that is to say, dominate) the market on a global scale.
>
> Finding a way to relate the two senses of the phrase "world cinema," and to balance the competing claims of the global cinema institution and the many different cinemas which exist throughout the world, has been the biggest single challenge in planning and putting together this book.

Given the sheer diversity of film industries and cultures included in the category of world cinema, Nowell-Smith (1996, pxx) concedes that "it would be foolish or arrogant, or both, for any one person to attempt to encompass the entire history of cinema single-handed." Yet the term remains very appealing for many film scholars and filmmakers who seek to bring diverse cinemas from around the world under a single convenient banner. For instance, the World Cinema Foundation, created recently in 2007 by the noted Hollywood director Martin Scorsese, proclaims on its website that it is "dedicated to the preservation and restoration of neglected films around the world." The Foundation's mission statement elaborates on its main objectives as follows:

> The World Cinema Foundation (WCF) is a non-profit organization dedicated to preserving and restoring neglected films from around the world—in particular, those countries lacking the financial and technical ability to do so.
>
> Established by Martin Scorsese, the Foundation supports and encourages preservation efforts to save the worldwide patrimony of films, ensuring that they are preserved, seen and shared. Its goal is to defend the body and spirit of cinema in the belief that preserving works of the past can encourage future generations to treat film as a universal form of expression.
>
> Cinema is an international language, an international art, but, above all, it is a source of enlightenment. There are wonderful, remarkable films, past and present, from Mexico, Eastern Europe, the Middle East, Latin America and Central Asia that deserve to be known and seen. Martin Scorsese has created the World Cinema Foundation with the specific purpose of calling attention to the global cause of film preservation. Composed of filmmakers from every continent, the World Cinema Foundation breathes life into the idea that when a cultural patrimony is lost, no matter how small or supposedly "marginal" the country might be, we are all poorer for it.
>
> http://worldcinemafoundation.net/mission/

In the "Message from the Chairman" that appears on the homepage of the foundation's website, Scorsese explains that the World Cinema Foundation is an extension of the Film Foundation he created in 1990 to preserve American cinema. Elaborating on the connection between the two foundations and, by extension, on the relationship between "American" and "world" cinemas, Scorsese writes:

> The World Cinema Foundation is a natural expansion of my love for movies. Seventeen years ago, together with my fellow filmmakers, we created The Film Foundation to help preserve American cinema. Much has been accomplished and much work remains to be done, but The Film Foundation has created a base upon which we can build. There is now, I believe, a film preservation consciousness.
>
> The World Cinema Foundation is being created to help developing countries preserve their cinematic treasures. We want to help strengthen and support the work of international archives, and provide a resource for those countries lacking the archival and technical facilities to do the work themselves.
>
> "Message from the Chairman," http://worldcinemafoundation.net/mission/

I have quoted extensively from the mission statement of the World Cinema Foundation and Scorsese's message as the chairman of the board not because I consider it to be a unique organization engaged in the preservation of "cinematic treasures" from around the world, or because I find it singularly guilty of trying to do too much with the category of "world cinema." There are many other organizations and individuals who have enthusiastically—and often uncritically—embraced the term "world cinema," and the World Cinema Fund, established in 2004 as a joint venture between the German Federal Cultural Foundation (Kulturstiftung des Bundes) and the Berlin International Film Festival, is a good example of another leading institution in this enterprise that predates Scorsese's foundation by a good three years.[2] However, I want to closely read the mission statement of the World Cinema Foundation and Scorsese's message to the readers because these texts highlight a prevailing and dominant view in Hollywood-based film industries and Hollywood-centered film studies that demarcates the world of cinema into a very neat binary of "American cinema" and "other" cinemas (from the Third World, developing countries, non-Western nations, etc.). Moreover, in this binary framework of American cinema versus "other" cinemas, the category of "world cinema" is also erroneously conflated with "national cinemas," thus marginalizing the identity and autonomy of "regional cinemas."

As Stephen Crofts (1993) points out, the term "national cinema" has traditionally been used to refer to cinematic cultures and industries that are not Hollywood, and emerge from countries other than the US. In his extensive survey of "national cinemas," Crofts finds that the term encompasses a diverse range of production, distribution, consumption, and regulation practices in different parts of the world. However, he suggests that in more general terms, there are at least seven varieties of "national cinema," which he lists as follows:

1 cinemas which differ from Hollywood, but do not compete directly, by targeting a distinct, specialist market sector;
2 those which differ, do not compete directly, *but* do directly *critique* Hollywood;
3 European and Third World entertainment cinemas which struggle against Hollywood with limited or no success;
4 cinemas which ignore Hollywood, an accomplishment managed by few;
5 Anglophone cinemas which try to beat Hollywood at its own game;
6 cinemas which work within a wholly state-controlled and often substantially state-subsidized industry; and
7 regional or national cinemas whose culture and/or language take their distance from the nation states which enclose them (Crofts 1993, p50).

While one may quibble with Crofts on the numbers and varieties of national cinemas found around the world, it is important to recognize, as he does, that in this Western-centric system of categorizing film productions, Hollywood is almost never described as a "national cinema." For example, on the Filmmakers Board of the aforementioned World Cinema Foundation, various well-known filmmakers from around the world are listed as representatives of different "national" cinemas: Fatih Akin (Germany/Turkey), Souleymane Cissé (Mali), Guillermo Del Toro (Mexico), Stephen Frears (UK), Alejandro Gonzales Inarritu (Mexico), Abbas Kiarostami (Iran), Deepa Mehta (India), Ermanno Olmi (Italy), Raoul Peck (Haiti), Cristi Puiu (Romania), Walter Salles (Brazil), Abderrahmane Sissako (Mauritania), Elia Suleiman (Palestine), Bertrand Tavernier (France), Wim Wenders (Germany), Wong Kar Wai and Tian Zhuangzhuang (China).

Although Scorsese is listed as representing the US on the Filmmakers Board (in addition to being listed as the chairman of the board of directors on the executive board), it is important to note that the World Cinema Foundation does not deal with preservation of films from the US—that is the responsibility of the Film Foundation (also headed by Scorsese). In thus demarcating "American" cinema from "other" cinemas and by conflating the latter with the "national cinemas" of countries beyond the US, the World Cinema Foundation—with all its noble intentions of preserving neglected cinemas around the world—continues to perpetuate trait-based geographies of colonialism (West versus the rest), Cold War ideologies (First, Second, Third Worlds), and the world capitalist system (Western "core" and non-Western "peripheries").

During recent years, some media scholars have used the term "transnational cinema" in place of "world cinema" to address the many limitations of the latter as it has been traditionally used in film studies. For example, an academic journal entitled *Transnational Cinemas* was recently launched in 2010 to address what the editors describe as "a shift in global film cultures." In the journal's mission statement, the editors elaborate on the global "shift" in the following terms:

> Transnational Cinemas has emerged in response to a shift in global film cultures and how we understand them. Dynamic new industrial and textual practices are being established throughout the world and the academic community is responding. Transnational Cinemas aims to break down traditional geographical divisions and welcomes submissions that reflect the changing nature of global filmmaking.[3]

However, as Will Higbee and Song Hwee Lim (2010) point out in an introductory essay for the first issue of the journal *Transnational Cinemas*, the term "transnational" is not an entirely unproblematic replacement for what has rather problematically been referred to as "world cinema" or "national cinemas." For instance, the term "transnational cinema" still privileges the space of the nation state (or at least the need/desire to overcome the nation-state framework) and thus marginalizes the rich diversity of cultural and geographic descriptors that define cinema in more global terms. According to Higbee and Lim (2010), there are three major theoretical frameworks which have been used in film studies to interrogate the questions of "transnational cinema." The first, they argue, is an approach that focuses on a national/transnational binary, and posits the "national" as a "limiting" model. This approach, exemplified by the work of Higson (2000), seeks to invoke a subtler meaning of the term transnational by focusing on production, distribution, exhibition, and reception of films across national borders and beyond their "indigenous" locations. Higbee and Lim argue that the major drawback of this approach is that it could potentially obscure the political, economic, and ideological imbalances of power that exist across and within borders, and ignore politics of difference in issues of migration and diaspora which are central to transnational flows in the contemporary world.

Another perspective in transnational cinemas that Higbee and Lim (2010) outline is the study of diasporic, exilic, and postcolonial cinemas. This category of transnational cinemas, Higbee and Lim argue, challenges Western theories that present nationalism and national cinemas as stable, pre-given entities, and critiques the Eurocentric biases of such theoretical constructs by analyzing cinematic representations of cultural identity as fluid, dynamic, and hybrid. Examples of studies in this category include books by Naficy (2001), Marks (2000), and Enwezor (2007). Such studies, Higbee and Lim (2010, p10) argue, are focused "almost exclusively on exilic, diasporic or postcolonial film-makers working within the West and are keenly aware of power relations between centre/margin, insider/outsider, as well as the continual negotiation between the global and local that often extends beyond the host/home binary in transnational or diasporic cinema." For Higbee and Lim, a potential limitation of the third approach is that diasporic, exilic, and postcolonial "transnational" cinemas are often marginal to the dominant film industries and cultures of which they are part, and thus their impact upon the mainstream is often difficult to evaluate.

While Higbee and Lim (2010) are acutely aware of the Eurocentric biases of Western theories of nationalism and, by extension, national cinemas, in their analysis of diasporic, exilic, and postcolonial cinemas they seem surprisingly uninterested in exploring the implications of such biases in their discussion of the category of "regional cinema" as transnational cinema. Higbee and Lim argue that the study of regional cinema as transnational cinema requires us to look at the transnational in terms of "regions" that share a common cultural heritage and/or geopolitical boundaries. As examples of this type of analysis, they cite Sheldon Hsiao-peng Lu's (1997) edited volume on transnational Chinese cinemas; Andrew Nestingen and Trevor Elkington's (2005) collection on transnational Nordic cinema; and Tim Bergfelder, Sue Harris, and Sarah Street's (2007) book on set design in European cinema of the 1930s. However, Higbee and Lim (2010) wonder if the term transnational is even an appropriate one to describe these "regional" cinemas. Instead, they propose that we speak instead of a "supranational Chinese" cinema, a "regional Nordic" cinema, and a "pan-European" cinema.

While I do not entirely disagree with Higbee and Lim's (2010) assessment that "transnational" may be an inadequate term to define these "regional cinemas," there is a fundamental problem with their understanding of "regions" purely in terms of "traits" such as common cultural heritage and shared geographic boundaries. By categorizing these "regional" cinemas as "supranational Chinese," "regional Nordic," and "pan-European," Higbee and Lim seem to reduce the category of the "region" to trait geographies of cultural heritage and physical boundaries. In doing so, they propagate what Appadurai (1999, p232) has identified as the dominant paradigm in the study of areas and regions that takes "a particular configuration of apparent stabilities for permanent associations between space, territory and cultural organization." For example, when histories of Indian cinema are written in terms of the perceived stability of traits such as language and physical borders, "regions" traditionally map onto linguistic states within the nation state and are thus relegated to the sub-national status alone even though such "regions" have historically been supra-, trans-, and pan-national as well.

The Trait Geographies of Indian Cinema

Cinema in India has been a very global and a rather hybrid phenomenon from the very beginning and thus defies neat categorizations. These hybrid beginnings make the task of historians more difficult, particularly for those who seek to write a linear narrative of its historical developments in terms of some perceived stability of traits such as cultural heritage, nationalist origins, physical boundaries, and so on. Yet, as Jigna Desai and Rajinder Dudhra (2008) argue, the history and historiography of Indian cinema has been driven by a nationalist desire to create coherent narratives which are internally consistent with the dominant ideologies of the modern nation state. Thus, these nationalist histories often seek to trace linear, progressive narratives of individual films, technological innovations, directors/auteurs, film studios, and performers as evidence of the emergence, development, and consolidation of a modern nation state and a modern nationalist cinema.

Much has been written on this topic, and it would be futile to provide an exhaustive account of this nationalist history and historiography on Indian cinema. Instead, I will only provide a brief overview of the traditional narratives of nationalist history in Indian cinema as a way to lay some groundwork for the critique of its linear teleology that follows. As chronicled by Ashish Rajadhyaksha and Paul Willemen (1999) in the *Encyclopedia of Indian Cinema*—and repeatedly highlighted in many history books and online sites introducing Bollywood or Indian cinema—unlike many other technological innovations of the colonial era, cinema arrived in India almost at the same time as it did in the imperial capitals of Europe. The first ever screening in India of six silent shorts films by the Lumière cameraman Maurius Sestier was held at the Watson's Hotel in Bombay on July 7, 1896. A year later, Calcutta and Madras held their first film screenings, and in Bombay Clifton and Co. begin daily screenings at their Medows Street Photography Studio. In 1898, two Italians, Colorello and Cornaglia, showed films in tents at the Azad Maidan grounds in Bombay. During the same year, Hiralal Sen started making films in India, and Amritlal Bose screened a set of "actualities" and "fakes" at the Star Theater in Calcutta. The Warwick Trading Company commissioned a newsreel, *Panorama of Calcutta.* Other films made in 1898 include *Pooa Races* and *Train Arriving at Churchgate Station* (by Andersonscopograph). Harishchandra. S. Bhatavdekar, also known as Save Dada, who made two short films—*Wrestlers* and *Man and Monkey*—in 1899, is considered the first Indian filmmaker by some film historians (Chatterjee 2000).

In 1900, F. B. Thanawala established Grand Newsreels and released films such as *Splendid New View of Bombay* and *Taboot Procession.* During the same year, news footage of the Boer War was released at Novelty Cinema in Bombay, and Major Marwick established a cinema in Madras. In 1901 Hiralal Sen's Royal Bioscope filmed extracts from plays and began film exhibitions with a commercial theater in Calcutta. Also in 1901, Bhatavdekar filmed the landing of Sir M. M. Bhownuggree and the return to India of Sir Wrangler Mr. R. P. Paranjpye from Cambridge University. Bhatavdekar and the American Biograph Company filmed Lord Curzon's Delhi Durbar in 1903 to mark the enthronement of King Edward VII (Rajadhyaksha and Willemen, 1999).

In 1902, J. F. Madan launched a bioscope show at Calcutta Maidan in a tent. Madan was perhaps the first Indian businessman to consider the commercial possibilities of distributing and exhibiting films on a large scale on the subcontinent. In 1905, Madan became a film producer when he produced Jyotish Sarkar's film of a protest rally against the partition of India and Pakistan. Rajadhyaksha and Willemen (1999, p17) write that by the early 1900s, Madan had laid the foundations of "a massive exhibition and distribution empire which dominated silent India, Burmese and Sri Lankan cinemas," and Madan's Elphinstone Bioscope Co. was the dominant producer of Indian films by 1906. In 1907 Madan built Calcutta's first movie theater, Elphinstone Picture Palace, and by 1927 he owned a chain of 85 theaters in India, Burma (now Myanmar), and Sri Lanka (Chatterjee 2000).

Pundalik, the story of a Hindu saint from Maharashtra, was the first feature film made in India and released on May 8, 1912 at the Coronation Cinematograph in Bombay. However, the status of *Pundalik* as the first "indigenous" film is disputed by some nationalist historians because it was made by P. R. Tipnis with British collaboration. Instead, history books bestow the honor of making the first indigenous feature film in India to Dhundiraj Govind Phalke (also known as Dada Phalke), whose first feature film *Raja Harishchandra* was released in 1913. *Raja Harishchandra* was a mythological film based on a short segment of the Hindu religious epic, the Mahabharata, and the story goes that Phalke was inspired to the film when he saw the film *Life of Christ* in a theater in Bombay in 1910. Recalling the incident, Phalke (1917) later wrote:

> When the *Life of Christ* was rolling fast before my physical eyes, I was mentally visualizing the Gods, Shri Krishna, Shri Ramachandra, their Gokul and Ayodhya. ... I felt my imagination taking shape on the screen. Could this really happen? Could we the sons of India, ever be able to see Indian images on the screen? The whole night passed in this mental agony.
>
> Cited in Ganti (2004, p9)

By taking his inspiration from Western cinematic narratives and religious traditions, and recasting traditional Hindu stories from religious epics, classical and folk theaters, art and literature for the new medium of cinema, Phalke helped to create a new genre of Hindu mythological films that captivated audiences in India. For this achievement, Phalke has been hailed as the "Father of Indian cinema" in many nationalist accounts. Even though Phalke was a "regional" filmmaker in the regional cinema of Marathi, this fact is rarely mentioned in many nationalist narratives of Indian cinema. Similarly, the fact that many of the early pioneers such as Hiralal Sen, Amrit Bose, Harishchandra. S. Bhatavdekar, and J. F. Madan were all "regional" players is also ignored within histories of Indian cinema, which are written from a nationalist perspective, or from the perspective of Bollywood as India's "national" cinema.

In *Bollyworld: Indian Cinema through a Transnational Lens*, Raminder Kaur and Ajay Sinha (2005) critique the nationalist teleologies in the historiography of Indian cinema for eclipsing not only the many regional variations but also the international connections that Phalke and his contemporaries demonstrated in their pioneering works. Moreover, Kaur and Sinha argue, about 90 percent of films screened in India during the early years were foreign films, and so Phalke's Hindu mythologicals did not reach or have an effect on all parts and all peoples of India. For example, they argue that many Muslims in India were reluctant to watch mythological movies about Hindu deities, and even Hindu audiences did not watch only the mythological genre because foreign films were more widely distributed in India at the time.

Kaur and Sinha argue that the arrival of sound technology during the 1930s enabled the production of films in several Indian languages, and paved the way for the rapid development of regional film industries across the subcontinent. The first Indian "talkie" released in 1931 was a Hindi-Urdu film, *Alam Ara*, based on an Arabian Nights fable. Sound feature films in three other Indian languages were also released in 1931: *Kalidasa* in Tamil, *Bhakta Prahalada* in Telugu, and *Jamai Sasthi* in Bengali. In 1932, the East India Film Company in Calcutta began producing films in Bengali, Tamil, and Telugu. During the same year, the first sound features in Gujarati (*Narasinh Mehta*) and Marathi (*Sant Tukaram*) were released (Rajadhyaksha and Willemen 1999). Since then, hundreds of films in different genres, on a variety of themes, in many different languages have been produced every year in India.

During the early years of sound cinema in India, the largest number of feature films were the Hindi-Urdu films made in Mumbai, and they were the most widely circulated, particularly in northern parts of the subcontinent and into other countries in Asia, the Middle East, and Africa. After India's independence from British rule in 1947, the Hindi language films produced in the Bombay-based film industry continued to dominate both in terms of numbers and popular appeal as pan-Indian "formula" films—particularly since Hindi is the official national language of the modern Indian nation state.

Until the late 1950s, Hindi cinema accounted for one third to half of all the feature films made in India each year. However, by the 1960s, Tamil and Telugu cinemas combined produced more films per year than the Hindi film industry, and in 1979 both Tamil and Telugu films outnumbered the Hindi feature films made during that year (the numbers for 1979 were 139 Tamil films, 131 Telugu films, and 113 Hindi films). By the 1980s, Tamil, Telugu, and Hindi film industries were all making approximately the same numbers of films each year. However, when numbers for Tamil and Telugu films are combined with the numbers for Kannada and Malayalam films, South Indian films account for almost 70 percent of the feature films made in India every year.[4] The feature films made in regional languages such as Tamil, Telugu, or Kannada can also be considered "formula" films like the Hindi films made in Mumbai. Not unlike their Hindi counterparts in Bollywood, the "formula" films in Tamil, Telugu, and Kannada are multi-star productions with big budgets, extravagant sets, exaggerated narratives, and elaborate song-and-dance sequences. Since the 1990s, several South Indian films have been made with bigger budgets, on a much larger scale, and with greater technical sophistication than many of their more celebrated counterparts from Bollywood (Pillai and Tilak 1995).

In both Hindi and other Indian language-based film industries, there are some filmmakers who define themselves by rejecting the formula-based conventions of commercial cinema in India.

These filmmakers and their films are usually classified as "art cinema" or "new cinema." Perhaps the best-known Indian filmmaker in the "art cinema" movement is the Bengali filmmaker Satyajit Ray. Ray's first film, *Pather Panchali*, won a prize at the Cannes film festival in 1955, and he became the representative of Indian cinema for many international audiences of "world cinema" who were largely unfamiliar with or uninterested in the commercial cinemas of India.

In contrast to commercial cinema in India, Ray's films are marked by a restrained and lyrical neo-realism, without stars, elaborate sets, or song-and-dance sequences. Ray's films are also known for their critical examination of social tensions between tradition/modernity, old/new, and rural/urban in Indian society. A third type of cinema called "Middle cinema" has also existed in India since the 1960s and 1970s. Middle cinema, as its name suggests, represents one of the most creative attempts in India to bridge the gap between "art" and "commercial" cinemas (Vasudev 1986).

A fourth type of cinema that has emerged since the 1990s is called diasporic cinema, or NRI (Non-Resident Indian) cinema. These films foreground diasporic desires as the central themes of the narratives. Bollywood superstar Shah Rukh Khan is the undisputed king of NRI cinema in India and in the Indian diaspora. Sometimes the term NRI cinema or diaspora cinema is also used to refer to films made by people of Indian origin in Canada, the UK, the US, and other parts of the world. The best-known among these filmmakers are Mira Nair, Deepa Mehta, and Gurinder Chadda, who have made films such as *Monsoon Wedding*, *Fire*, and *Bend It Like Beckham*.

Some scholars prefer to use the term "Bollywood" to define Hindi films that target NRIs, and/or films made by diasporic Indians since the 1990s, as a way to demarcate the current trends of "Bollywoodization" in Indian cinema from other cinemas (Rajadhyaksha 2003). Even if we were to use the term "Bollywood" in a more expansive way to describe all commercial Hindi films ever made in Bombay/Mumbai, it still represents a very particular segment of the Indian film industry, which consists of commercial cinema, art cinema, and parallel cinema in Hindi as well as commercial, art, and parallel cinemas in a variety of regional languages in India. To return to the main argument of this chapter, there is a need to distinguish between the hype about globalization of Bollywood in recent years and the more complex global histories and geographies of Indian cinema. I argue that the term "Bollywood" represents only one wave of globalization in India cinema.

Granted, it is a very powerful and dominant-hegemonic wave, but as I discuss in the following section, there are many other ways in which regional cinemas in India are, and have been, going global.

From Trait Geographies to Process Geographies

One of the most interesting aspects of globalization in Indian cinema is the way in which regional cinemas of India are going global and transforming their "sub-national" status based on trait geographies of language and location. In this context, when we shift the framework of analysis for regional cinemas from trait geographies of linguistic, cultural, and physical borders to process geographies of "trade, travel, pilgrimage, warfare, proselytization, colonization, exile and the like" (Appadurai 1999, p232), many regional cinemas—such as South Indian regional cinemas—appear to be at once supra-national, transnational, pan-national, and sub-national. The recent emergence of Ramoji Film City in Hyderabad and Innovative Film City in Bangaluru is a case in point.

Ramoji Film City (RFC) is the largest, most comprehensive, and professionally planned film production center in the world. According to the *Guinness Book of World Records*, it has surpassed Hollywood's Universal Studio in both size and the range of media facilities offered. RFC is the dream project of Cherukuri Ramoji Rao, the owner of the Eenadu media group in the South Indian state of Andhra Pradesh. The Eenadu group is one of the largest media conglomerates in South India, and Ramoji Rao's business empire consists of several English and Telugu-language periodicals, including the widely read newspaper *Eenadu*; a multi-lingual satellite television network, ETV; a film distribution banner, Ushakiron Movies, and a financial services group, Margadarshi.

A veteran producer and director in the Hyderabad-based Telugu film industry—or Tollywood, as some fans call it—Ramoji Rao recognized that filmmaking in India is a rather tedious and expensive venture because the production calendar can often be interrupted by unpredictable weather, unreliable electricity, inappropriate set designs, and inflexible star schedules. Realizing that a more reliable environment and a well-organized production schedule could dramatically bring down the time, the cost, and the resources required to make a film in India, Rao set about the ambitious task of developing a one-stop studio in Hyderabad.

Following the success of RFC, Innovative Film City (IFC) was launched on January 18, 2008 in Bidadi, which is about 50 kilometers from the city of Bengaluru. IFC is part of the Innovative Group, which runs a multiplex cinema business along with media production and entertainment and leisure activities in the State of Karnataka in India. In the state of Tamil Nadu, plans are afoot to transform the erstwhile MGR Film City into a one-stop pre-production, production, and post-production facility along the lines of RFC and IFC (Narasimhan 2008). Similar proposals have also been mooted in other states such as West Bengal and Bihar, but their proponents have had little success so far.

Since its opening in 1997, six foreign films and over 500 Indian films in languages such as English, Hindi, Kannada, Telugu, and Tamil have been produced at RFC. *Maa Nannaki Pelli* (*My Father's Wedding*) was the first Telugu film shot entirely at RFC in 1997. The first Hindi film shot in RFC was *Bade Miyan Chote Miyan* (*Big Guy, Little Guy*) in 1998, starring Bollywood superstars Amitabh Bachchan and Govinda. Several international films, such as *Centipede* (2004), directed by Gregory Gieras, *Crocodile 2: Death Swamp* (2002), directed by Gary Jones, *Nightfall* (2000), directed by Roger Corman, *Panic, Quicksand* (2002), directed by Sam Firstenberg, and *In the Shadow of the Cobra* (2004), directed by Ted Nicoloau, have also been produced in RFC. On the other hand, IFC was inaugurated more recently in 2008 and has a much smaller portfolio of films that have been made there. However, some filmmakers from Kannada cinema have used the production facilities at IFC in the past two years, and Innovative Group plans to promote the film city as a tourist destination and a production center on a much larger scale in the coming years.

What makes the concept of turnkey productions at film cities such as RFC and IFC rather unique in Indian cinema is that for the first time filmmakers from anywhere in the world can make an entire film from pre-production to post-production in a one-stop studio that provides multiple outdoor locales and diverse indoor settings. In addition to being state-of-the-art media production centers, RFC and IFC are major tourist attractions that provide visitors access to a variety of picturesque gardens, entertainment parks, and tours of production studios.

Among the tourist attractions at IFC are a mirror maze, a dino park, a haunted mansion, and a wax museum which displays life-like models of prominent real and fictional figures such as Mahatma Gandhi, Bill Clinton, Harry Potter, Sleeping Beauty, and Madonna. Funplex is a digital media park which includes virtual sports, a 360-degree simulator, and a 4-D cinema theater that combines 3-D movies with effects such as moving chairs, wind blowing, water spraying, leg ticklers, virtual fog, rain, lightning, air bubbles, and smells. Aqua Kingdom is a 2.2 acre waterpark that seeks to recreate the ambience of being at a beach. IFC is also home to Ripley's Believe-It-Or-Not Museum and a Jurassic Park-type Fossil Museum with a 60 foot high display of a T-Rex skeleton along with skeletal displays of many other dinosaurs.

Similarly, many of the attractions at RFC are clearly recognizable to tourists since they are, rather stereotypically, marked as "North Indian," "South Indian," "Western," or "Japanese." In the "North India" section of RFC, for instance, there is an exact replica of the ancient Fort City of Fatehpur Sikri near Agra. There is a replica of the Taj Mahal, too, but it can be easily converted to resemble a mosque with a backdrop of a crescent moon at night. Similarly, a building with a front elevation that resembles the Golden Temple in Amristar, Punjab, has a rear elevation that looks like a shopping center. In RFC, one can travel from Mumbai or Calcutta to London or New York in a few minutes. With a few modifications, the façades of high-rise buildings at the film city can be made to resemble a busy street in London, or an upscale neighborhood in Mumbai, or a scenic ski resort in the Swiss Alps.

Of course, the synthetic reproductions of prominent cultural landmarks and popular tourist attractions at RFC and IFC are not "real" places in the traditional sense of the word. Instead, the creations at RFC must be understood in terms of what Jean Baudrillard (2000) has defined as hyper-real constructs—simulations of reality which appear as "real" as the real things themselves. In this postmodern sense of simulacrum that Baudrillard speaks of, the cultural creations and recreations at RFC and IFC are prominent examples of hybrid mediascapes that can be used by filmmakers and tourists from all over the world.

Mediascape is a term coined by Arjun Appadurai (1997) to describe the diversity of media images that is constantly produced and reproduced by new media technologies in a rapidly globalizing cultural economy. Appadurai's formulation of mediascapes provides a useful framework to examine how the production practices at RFC have emerged at the intersection of global, national, and local flows of media and popular culture in South India. By foregrounding the hybrid mediascapes being produced at RFC and IFC, this chapter underscores the creative ways in which film cities such as RFC and IFC in South India are mapping a new process geography of film production and consumption in Indian cinema. In conclusion, I argue that film cities such as RFC and IFC are very good case studies for understanding the changing realities of the global entertainment industry where connections between older trait-based definitions of place are being re-imagined in terms of newer process-oriented meanings of mobility, travel, tourism, and the like.

Notes

1 For general definitions of "world cinema" see, for instance, Badley, Palmer and Schneider (2006); Chaudhuri (2005); Gazetas (2008); Hill and Gibson (2000). The almost complete obscurity and marginalization of regional cinemas of India in this category of films is also evident in these introductory texts on the topic.

2 The World Cinema Fund website can be found at http://www.berlinale.de/en/das_festival/world_cinema_fund/wcf_profil/index.html.

3 The mission statement is available on the journal's website at http://www.intellectbooks.co.uk/journals/view-Journal,id=183/.

4 These numbers are compiled by Rajadhyaksha and Willemen (1999, pp30–32) from the official figures of censored films published by the Central Board of Film Certification in India. According to Rajadhyaksha and Willemen, these numbers include several dubbed films, and multilingual films are usually counted as distinct and independent productions for each language.

References

Appadurai, A. (1997) *Modernity at Large*. Minneapolis, MN: University of Minnesota Press.

Appadurai, A. (1999) "Globalization and the research imagination," *International Social Science Journal*, 51(160): 229–238.

Badley, L., B. R. Palmer, and S. J. Schneider (2006) *Traditions in World Cinema*. New Brunswick, NJ: Rutgers University Press.

Baudrillard, J. (2000) *Simulacra and Simulation*. Ann Arbor, MI: University of Michigan Press.

BBC News (1999) "Bollywood star tops poll," BBC News, http://news.bbc.co.uk/2/hi/entertainment/381017.stm, accessed 26 January 2000.

Bergfelder, T., S. Harris, and S. Street (2007) *Film Architecture and the Transnational Imagination: Set Design in 1930s European Cinema*. Amsterdam: Amsterdam University Press.

Chatterjee, S. (2000) "Cinema in India," in D. Hoiberg and I. Ramchandani (eds) *Students' Britannica India*, Volume 6. Encyclopedia Britannica, 97–104.

Chaudhuri, S. (2005) *Contemporary World Cinema: Europe, the Middle East, East Asia and South Asia*. Edinburgh: Edinburgh University Press.

Crofts, S. (1993) "Reconceptualizing national cinema/s," *Quarterly Review of Film & Video*, 14(3): 49–67.

Desai, J. and R. Dudhra (2008) "The essential Bollywood," in J. Desai and R. Dudhra (eds) *The Bollywood Reader*. Milton Keyes, UK: Open University Press, 1–17.

Enwezor, O. (2007) "Coalition building: Black audio film collective and transnational post-colonialism," in K. Eshun and A. Sagar (eds) *The Ghosts of Songs: The Film Art of the Black Audio Film Collective*. Liverpool: Liverpool University Press, 106–129.

Ganti, T. (2004) *Bollywood: A Guidebook to Popular Indian Cinema*. London and New York: Taylor & Francis.

Gazetas, A. (2008) *An Introduction to World Cinema*. Jefferson, NC: McFarland.

Higbee, W. and S. H. Lim (2010) "Concepts of transnational cinema: Towards a critical transnationalism in film studies," *Transnational Cinema*, 1(1): 7–21.

Higson, A. (2000) "The limiting imagination of national cinema," in M. Hjort and S. MacKenzie (eds) *Cinema and Nation*. London: Routledge, 63–74.

Hill, J. and P. C. Gibson (2000) *World Cinema: Critical Approaches*. New York, NY: Oxford University Press.

Kaur, R. and A. Sinha (2005) *Bollyworld: Indian Popular Cinema through a Transnational Lens*. Thousand Oaks, CA: Sage Publications.

Lu, S. H. (ed) (1997) *Transnational Chinese Cinemas: Identity, Nationhood, Gender*. Honolulu: University of Hawaii Press.

Marks, L. (2000) *The Skin of the Film: Intercultural Cinema, Embodiment, and the Senses*. Durham and London: Duke University Press.

Mehta, S. (2005) "Welcome to Bollywood," *National Geographic*, February, 52–69, http://ngm.nationalgeographic.com/ngm/0502/feature3/, accessed 5 January 2006.

Naficy, H. (2001) *An Accented Cinema: Exilic and Diasporic Filmmaking*. Princeton and Oxford: Princeton University Press.

Narasimhan, T. E. (2008) "Movie park planned in city," *Business Standard*, 10 October, http://www.business-standard.com/india/news/movie-park-planned-in-city/336878/, accessed 15 March 2006.

Nestingen, A. and T. G. Elkington (eds) (2005) *Transnational Cinema in a Global North: Nordic Cinema in Transition*, Detroit, MI: Wayne State University Press.
Nowell-Smith, G. (1996) "General introduction," in G. Nowell-Smith (ed) *Oxford Dictionary of World Cinema*. Oxford: Oxford University Press, xix–xxii.
Perry, A. (2003) "The new face of film," *Time* (Asia edition), 20 October, http://www.time.com/time/asia/covers/501031027/story.html, accessed 5 January 2006.
Pillai, A. and S. G. Tilak (1995) "Bettering Bollywood," *India Today*, 15 March, pp186–190.
Rajadhyaksha, A. (2003) "The Bollywoodization of the Indian cinema: Cultural nationalism in a global arena," *Inter-Asia Cultural Studies*, 4(1), April: 25–39.
Rajadhyaksha, A. and P. Willemen (1999) *Encyclopedia of Indian Cinema*. London: Fitzroy Dearborn Publishers.
Thoraval, Y. (2001) *Cinemas of India (1896–2000)*. Delhi: South Asia Books.
Vasudev, A. (1986) *The New Indian Cinema*. Bombay: McMillan India Limited.
World Cinema Foundation (undated) "From the Chairman," World Cinema Foundation, http://worldcinemafoundation.net/, accessed 15 March 2010.

Chapter 6

Mobilizing Global Communication

For What and For Whom?

Karin G. Wilkins

Building from previous chapters in this volume exploring how global communication becomes manifest in the industries and experiences of our communities, in our next set of chapters we question the value of global communication. To begin this critical exposition, we begin by suggesting substantive interrogation of our definitions of "global" as well as "communication."

Asserting the term "global" over other frames of reference, such as "international," privileges communication flows that are less restricted along boundaries of national sovereignty. By not focusing on international boundaries, perhaps more pertinent in the regulation of print and some broadcast media, we can include transnational, transversal, and regional interests into our realm of inquiry. Moreover, moving beyond a categorization based on national boundaries opens up space for dialogue concerning political communities not grounded in national territory. Our underlying interest in power differences moves us to an alternative geometry, in which access to resources through social, cultural, financial, and political capital matters more than geographical positioning within the nation state (Shah and Wilkins 2004).

Our focus on "communication" over "media" positions our analyses within social and political contexts, grounded in economic foundations. We are interested not just in the particular capacity of the communications technology or the content, but more so in how we as a global community relate to media in a broader communication environment. In Chapter 10, for example, de Madariaga expands upon how journalists use digital media in their professional work, while in Chapter 7, Enghel explores organizational parameters around decisions to use communication in social change projects. In Chapter 8, Brooten also positions her attention to media within the work of social movements, just as Chakravartty (Chapter 9) considers the broad societal implications of information communication technology (ICT) industry practices.

Two trends are worth noting for the purposes of this and subsequent chapters. First, global communications technologies have been expanding, even if slowly, in terms of their technological capacities as well as people's access to them (see Chapter 1 in this volume). Who has access to communication technologies is largely restricted to an elite educated group. Many communication technologies transcend or traverse national boundaries, constituting a broad domain of global communication, such as satellite television, though the Internet is often touted as a central component of this global network, offering the possibility for collective mobilization and communication across sovereign states. The growth of Internet usage from 2000 to 2010 has been dramatic, having increased by 445 percent, most notably in the African, Middle Eastern, and Latin American regions; yet another look at these same statistics demonstrates that still fewer than one third of the world's population (28.7 percent of approximately 6,845,609,960 people) use the Internet (Internet World Stats 2010). Within North America, Oceania/Australia, and Europe, more than half of the documented populations (though only 77 percent in North America) use the Internet, but the proportion of users goes down dramatically in other regions.

The second trend worth noting is that many global conditions, particularly poverty and hunger, are getting worse rather than better. The first Millennium Development Goal agreed upon by the United Nations (UN) and other development agencies aims to eradicate extreme hunger and debilitating poverty, yet their own assessment of this goal indicates that between 1990 and 2009 we have more rather than fewer people who are hungry, living in poverty, or accruing insufficient wages over time (United Nations 2010). So even though the technological capacities of global communications may be increasing, we have few benefits accruing to the vast majority of the global population, most of whom still suffer from inadequate water supplies and healthcare. This recognition inspires our questioning of who actually benefits from global communication.

We do recognize the potential for communication to offer strategic value to groups engaged in social change and in resistance. With the ability to mobilize action quickly, and for purposes antithetical to corporate and government interests, global communication, when accessible, allows producers of communicative content to project perspectives that may differ from those offered through mainstream media resonating with dominant hegemonic interests. But there are fundamental concerns here with these basic premises of access and control.

If the Internet and other interactive digital technologies are to become more than merely broadcasters of information created by and for dominant agencies, there must be the potential for open content creation and distribution. Government and corporate control may limit the potential for global communication to serve public interests. These issues are well illustrated in Chapter 9 in this volume. For global communication to benefit people who do not reap the rewards of corporate profit or government regulation, policies need to structure communications industries in ways that increase access and permit user-generated content. These policies can be seen in how journalistic practices and ICT industries are governed.

Expanding our scope of interest to include a variety of organizations and communities mobilizing global communication parallels recent shifts in conceptualizing the use of communication in the development industry to reinforce hegemonic interests of wealthy bilateral donors, toward including attention to social movements' strategies, toward engaging in participatory processes and activist resistance. Increasing attention to social justice in the field of development communication highlights our interest in not just assessing the effects of communication campaigns, but understanding instead how the campaign itself becomes part of people's access to information and other resources in order to mobilize and act. At issue is the potential control over the production of communication about one's self, one's community, and one's vision of the world.

Our collective works in these chapters address the industry of ICTs, the practice of journalism, and the discourse of development and social movements, problematizing who benefits from their mobilization of global communication. Although these chapters offer serious critiques, they also presuppose a fundamental hope that by raising awareness of inequities, we can collectively work toward more just access to and use of information resources. Through the process of deconstruction, we can help contribute to the gradual reconstruction of a more equitable and fair world.

Advocacy Communication

In response to our concern that global communication needs to address how communication processes and technologies might work to benefit and not harm marginalized communities, I propose a framework for advocacy communication in a global context. Advocacy communication is situated within a broader literature of communication for social change, building from critical analyses of the development industry. This approach is distinct from social marketing in its emphasis on normative and structural change, while bridging entertainment education with media advocacy through proposing an integrated strategy that encompasses popular culture and news genres accessed through a variety of communicative modes. Prior to conceptualizing this framework, I briefly review a history of communication approaches to development and social change, considering the value of transitions as well as the unfortunate cooptation of key terms.

Development Communication

The proposed model for advocacy communication builds from a critical history of approaches in the fields of development, social change, and social movements, using strategic communication to promote a public good, such as improvements in conditions of health, agriculture, governance, and material welfare. Moreover, the model considers social movements' ways of addressing messages and seeking access to mainstream media institutions (Ryan 1991).

The model derives from a critical assessment of the potential utility as well as the limitations of social marketing, entertainment education, and media advocacy.

Historically, development communication can be seen as using communication *for* development, in terms of building media infrastructure, in addition to preparing strategic messages, as well as understanding communication *about* development, with the emergence of more critical considerations of discourse and praxis (Wilkins and Mody 2001). Based on a critical assessment of these approaches, advocacy communication is meant to offer a potentially constructive approach rather than festering in purely deconstructivist modes of analysis in explorations of communication about development. Moreover, this framework approaches communication as a site in which competing groups can challenge dominant interpretations and negotiate for resources allocated to social change. This critical approach posits communications as a process of asserting and contesting discourse within structures of power, rather than assuming a pluralist model of social change often dominating studies of communication for development.

Using communication *for* development has been the central focus of research, both in terms of academic as well as industry attention. Evaluations of social marketing and entertainment-education projects have drawn attention to the potential for media to change the knowledge, attitudes, and practices (KAP) of individuals, when sufficiently motivated to do so. While these projects have been somewhat varied in terms of their effects (though across a range of projects; Snyder 2002, estimates an average change in behavior at 10 percent), depending on the resonance of the issue with the targeted community, the quality of the project (Hornik 1988), the political commitment of agencies and communities, and the funding levels from donors, these approaches maintain their focus on individuals as a central focus for social change. Studies of communication about development critique these projects for their preoccupation with individuals as the key figure in the social change process at the expense of understanding cultural, political, and economic contexts that guide and constrain individual action (Wilkins 2000).

This critical approach to understanding the ways in which development approaches communicate assumptions about strategic social change also offers a useful guide for understanding a history of concepts in the field (Escobar 1995). Modernization approaches to development, not only prominent in the early years of development communication literature but also still present in the development discourse of prominent bilateral and multilateral donors, articulate a linear model of preconceived national transition from traditional to modern societies (Lerner 1958; Schramm 1963). Modern society is conceived as democratic and capitalist, while individuals are expected to propel this transition through their empathic projections from local to national alignments, through their exposure to media (particularly radio and news in Lerner's projections). This model has been critiqued for projecting Northern, Western models, particularly positioned within white, male, middle-class perspectives, as universal development trajectories, for emphasizing the nation as the critical venue and the individual as the central actor in the social change process, and for privileging science and technology over other approaches to knowledge production (Melkote and Steeves 2001).

Dependency critiques of the modernization approach recognized the importance of global contexts in understanding national development, and brought political interests and economic structures into focus (Schiller 1991). Privileging structural dimensions of development, dependency approaches were faulted, though, for not recognizing the agency of individual and collective actors outside of government and corporate agencies.

Asserting the potential of communities to engage development processes of their own, participatory approaches emerged as a critique to modernization for following too much of a top-down process, and to dependency for minimizing the ability of local groups to engage in decision-making (Jacobson and Servaes 1999). While modernization placed development within a national context and dependency in a global context, participatory approaches emphasized the role of communities as a central venue for collective action.

While "participation" was proposed initially in response to critiques of earlier approaches as a coherent and moral model for social change, over time the meaning of this term lost its initial value. Ambiguities grew over what participation signaled, as some argued that participatory approaches should be used as an ethical approach to process, while in practice the term was also applied to attempting to promote more efficient strategies in order to achieve preconceived outcomes (Huesca 2002). Similar to the process through which the term "empowerment" became coopted (proposed initially as a strong stance articulating an ability to control resources and decisions but over time becoming integrated within a right-wing mantra arguing that individuals did not require public spending on health or welfare if they were to be empowered), the term participation unfortunately became coopted to the extent that is not clear exactly what is meant when the word is used.

While academic approaches to development communication became increasingly entangled with attempts to promote participatory models of social change, despite what was happening in the practice of the development industry, scholars of social change began to incorporate political and sociological attention to the work of social movements (Huesca 2001). Integrating social movements, along with community and citizens' media (Rodriguez 2000), meant understanding social change not just within the purview of the development industry, but also situated within community action. Broadening the scope of this field to include social movements and community media corresponded with a gradual shift in rhetoric from "development" to "social change."

Communication for Social Change

Over time, academic attention to "social change" has begun to eclipse "development," critiqued for its historically limited theory, particularly when approached within modernization paradigms articulating individual actors as most powerful within linear models of change, as well as its limited practice, engaging hierarchical diffusion implemented to advance the political and economic interests of donors.

While approaches to social change within communication have not entirely discarded attention to the development industry's hierarchical development style, with research analyzing media effects of projects using communication for development goals, this shift signals a more comprehensive framework, integrating the work of social movements and activists, as well as critical attention to what these processes communicate about our assumptions concerning the nature of problems and appropriate solutions. Research now incorporates attention to the texts and forms of communication, as well as political and economic structures of production (Wilkins 2000).

This shift in rhetoric may be more a part of academic and non-governmental organization (NGO) communication discussion, though, than represent actual shift in practice or within the development industry of bilateral and multilateral donors. The division within the International Communication Association has changed its name from "development" to "social change," formerly connected with "intercultural" and subsequently with "global." The Rockefeller Foundation and their recently established Communication for Social Change Consortium, among others, have also articulated "social change" as its central focus (Gumucio-Dagron and Tufte 2006).

The term "development" remains prominent, though, in development institutions, some academic educational programs (such as Malmö University's graduate degree), in fields focusing on media development and ICT, and even within more sociologically and culturally oriented publications (e.g., Hemer and Tufte 2005). While, in the last ten years, about 10 academic book and article publications in English have referenced social change in connection with communication, more than 20 have used "development" in their titles. Also worth noting is that these titles are not mutually exclusive, with many articles using both development and social change, representing an intent to appeal to both sets of interests (and perhaps an inability to choose positions).

The gradual shift in terms is not linear, just as neither would a projected shift from a constructed traditional to modern life operate in a preconceived linear direction. Rather, the use of terms involves a dialogic process of negotiation as groups with, at times, competing and, other times, intersecting interests assert their agendas through their rhetoric and practice. While the use of the phrase social change has not completely replaced articulated interests in development, the approach has achieved just enough attention to warrant critical reflection.

Problematizing Social Change

The phrase "social change" brings together two terms meant to resonate with comprehensive approaches to strategic intervention designed to benefit the public good. The term "change" assumes that there is first a condition that can be observed as changing over time. Moreover, within this framework, change also implies that strategic intervention has the possibility of influencing this condition. Without further explanation, though, the term change can be potentially problematic since not all change involves conditions that benefit people. Some change can hurt, even kill, people and other living organisms. For this reason attention to "benefit" is proposed as preferable to the more ambivalent "change."

So if the term change requires clear conceptualization, the term "social" is potentially even more problematic. Social in this context tends to imply work within a human community which would be for the benefit of a public good. Defining that public good becomes difficult and often contested from different perspectives. This framework privileges social, while marginalizing political, economic, and cultural conditions. Political dimensions are considerably critical in the broader field, drawing our attention to media development, governance, civic engagement, and activist movements. Economic conditions matter in terms of recognizing the import of material conditions, poverty, and equity of distribution. Cultural identity also needs to be incorporated within a broader framework attesting to cultural production. Relying on a purely human venue for change also excludes potentially intriguing work in artificial intelligence, recognizing the communicative potential of communication with and among non-human technologies (Hamelink 2008).

Opening up this framework to include political, economic, cultural, and technological concerns with social conditions also needs to include attention to contexts. Instead of conceiving of development within circumscribed local communities or national boundaries rooted in territories, transnational spheres within global contexts should be articulated within a new "geometry" of development (Shah and Wilkins 2004). This emerging geometry connects those with power in what have typically been referred to as centers in relation to peripheries, regardless of spatial positioning. This geometric mapping highlights access to social, human, and financial capital as more critical in terms of defining quality of life than geographical positioning in bounded territories. It is not necessarily the nation in which one resides that matters most, but rather one's access to material, political, and social resources for healthy living.

It is time now to move past articulating our field in terms of "post-" and "re-," connected with development and developing. Given the ambiguities of terms and the broadening of conceptual approaches, we can begin to assert more comprehensive frameworks for strategic intervention. This approach should be comprehensive through granting interdisciplinary recognitions of multiple conditions, bringing attention to issues of power and control over the production of communication. This should be a critical approach, understanding what social change approaches communicate about people and processes, in order to advocate for a long-term, systemic transition that is more about justice and rights than about more ambivalent change.

A Critical Assessment of Communication *for* Social Change

Communication for social change refers to strategic communication processes and media projects designed to advance causes for the public good. Based on a critique of how identified models for social change work, I build a case for *advocacy communication* as a more comprehensive communicative approach to global social change.

Creating this framework requires careful consideration of how other strategic approaches, social marketing, entertainment education, and media advocacy have conceptualized communication for social change. These models of communication intervention represent ideal types described in the field, intersecting theoretical approaches with applied practice. It is recognized that these approaches are neither mutually exclusive in theoretical terms, nor completely distinct in implementing specific projects.

Building on a critical approach to understanding communication about social change, I will, in what follows, explore ideological assumptions implied within these frameworks in order to assess potential bridges across approaches based on limitations within them. Instead of assuming a pluralist model of social change, I will build on critical assumptions that power differences matter in terms of ability to assert interpretations and allocate resources.

Working from a critical approach concerning "communication about social change," I explore ideological assumptions implied within these frameworks. I begin with the assumption that social problems are not given, but selected and constructed by groups, communities, and organizations (Rakow 1989; Salmon 1989; Escobar 1995). How these problems become characterized constrains possible types of solutions. Framing world population as a bomb implies that subsequent intervention should be immediate in order to diffuse the situation; if, alternatively, the problem of population was to be framed as one of overcrowding on a boat, then perhaps redistribution of people might be an appropriate response. Framing a problem as a "war" on poverty or on drugs implies that there are heroes and enemies, and that a battle must be won despite the cost.

The groups and organizations creating these strategic interventions, then, have power to define both the problems as well as the appropriate solutions. The construction of issues and interventions carries with it assumptions about models of social change, either seen as advanced through individualist action in a pluralist society or otherwise as constrained by collective, normative, and structural conditions. For example, focusing on individuals means encouraging seat belt use rather than creating policies to regulate car manufacturing; quitting smoking rather than restricting smoking in public spaces or reducing subsidies to tobacco farmers; as well as boiling water and washing hands rather than improving plumbing and water supply.

The three strategies addressed—social marketing, entertainment education, and media advocacy—all employ intentional organized strategies toward specific non-commercial goals. Strategic communication projects share an assumption that media have the potential to influence social change, whether as a mechanism toward informing or persuading individuals, shifting normative climates, or encouraging policy change. These types of projects differ in other ways, though, such as in terms of the types of groups they address, as well as the nature of the social change process assumed. For example, social marketing targets individual consumers to change their behavior, whereas media advocacy directs attention toward policy-makers who have the potential to change structural conditions. The strategies also differ in their theories of how media might influence individuals and society.

Social Marketing

Social marketing employs media with the purpose of attempting to persuade individuals to change their behavior, following a hierarchical path beginning with exposure to a message, leading through cognitive steps in comprehension, knowledge acquisition, and attitudinal and then behavioral change. According to conventional models of this approach: a *product* is *positioned* within competing ideas or things through the *promotion* of a particular campaign implemented through a *place* (distribution channels), demanding a *price* (monetary, time, and opportunity costs) to consumers (Solomon 1989; Kotler et al. 2002). This approach is similar to advertising in that media are assumed to have an impact on individual audience members, but different in its underlying concern with social rather than commercial benefit.

The process of social change assumed is pluralist in nature, building on the notion that individuals need to determine personal decisions that may then cumulatively lead toward broader change. The media's role in this process is to provide persuasive messages that will inform and inspire audiences to change in accordance with the goal of the campaign. Whether projected through print or visual technologies, messages are disseminated hierarchically from an institutional center to a diffuse conglomerate of individual spectators.

In social marketing, the format tends to employ short, precise advertisement structures, incorporating messages determined by the institution designing the campaign. Each decision in constructing the message, such as the wording, language, source, tone, color, music, and more, falls within the domain of the organizations creating the campaign strategy. Messages tend to be channeled through more mainstream media industries in an attempt to reach target audiences.

Within a social marketing framework, audiences are "segmented" and "targeted" in order to position a message that may directly persuade this group. Audiences may be defined in a variety of ways, focusing on individual characteristics, such as demographic, psychological, or lifestyle characteristics. How that audience is defined implies a set of ideological assumptions about the nature of the group to which the problem is attributed. For example, defining a target audience for a social marketing campaign promoting contraception as "married women of reproductive age" legitimizes sexual activity and responsibility not only within a particular gender and age group, but also within a legal status of relationship. By implication, men and unmarried women are not part of the target group for this information.

Campaigns in social marketing tend to use the phrasing "consumer" to describe their target audiences (Wilkins 1999). This nomenclature focuses on the ability of individuals to purchase products, advocating a particular approach to social change that is advanced through participation in a market economy. One way to measure success of social marketing campaigns, along with a projected hierarchy of effects transitioning through knowledge and attitude shifts, is to assess purchasing as a behavioral decision in accordance with a campaign message. For example, consumers inspired by population, health, and nutritional communication campaigns purchase condoms, oral-rehydration solutions, and soybeans. In a social marketing project sponsored by the US Agency for International Development (USAID) in Bangladesh (these and other examples are reviewed in Rice and Atkin 2001), for example, contraception is sold to men in public markets rather than freely distributed, intended to enhance the perceived value of the product.

Entertainment Education

Entertainment education builds on a social marketing approach, attempting to use media campaigns strategically to change audience attitudes and behaviors, but is distinguished through its more normative approach and its more narrative format. Instead of projecting concise, explicit messages through an advertising model, entertainment education embeds campaign messages within broader popular cultural texts. Fictional programs on television and radio are typically employed, as well as popular songs.

The intention of this approach is to *educate* audiences by attracting them to *entertaining* cultural productions, thus increasing the potential for exposure and attentiveness to campaign messages. The goal of this approach remains within the KAP (knowledge, attitude, practice) hierarchy for most campaigns (Singhal and Rogers 1999). The process of social change still tends to build on more psychologically oriented theories, projecting individual audience members' role modeling as well as para-social interaction (Singhal and Rogers 1999). An emerging trend, however, in this approach suggests that more normative and structural concerns might be engaged (Usdin et al. 2000).

Strategic communication campaigns implemented in many countries have used the entertainment education approach through television and radio fictional programming. In addition, some communication projects, funded by USAID, have incorporated messages within popular music, such as the Tatiana and Johnny project in 1986 promoting abstinence and songs about "choices" in Nigeria sung by King Sunny Ade and Onyea Onwenu (Singhal and Rogers 1999).

Media Advocacy

Media advocacy engages media both as a resource toward social change and as an intermediate focus through strategic intervention (Wallack et al. 1999). In attempting to change policies and structures, organizations and communities engaged in advocacy attempt to attract both attention and sympathetic news coverage of their issues as a channel toward mobilizing public support as well as reaching decision-makers. Media sources are also used as a resource for mobilizing people in collective action.

To offer one contrast within the realm of health communication, social marketing campaigns might focus on encouraging individual women to initiate and prolong breastfeeding; in contrast, a media advocacy project, such as one supported by the Infant Formula Coalition Action (INFACT), organized a boycott of Nestlé in response to its aggressive marketing of infant formula in developing countries (reviewed in Rice and Atkin 2001).

Policy-makers and other critical decision-makers with the authority to create and adapt structural conditions serve as the ultimate audience for media advocacy. Media professionals, engaged in the creation of text addressing issues of concern to the media advocacy group, serve as the penultimate audience. Barker-Plummer's study (1995) of women's movements, for example, illustrates different ways in which organizations might approach the news media as a political resource in a dialogic relationship in which activists decide to work pragmatically or subversively with media professionals. Mediated channels also work as a force toward mobilizing more actively engaged participants, thus acting as a resource as well as a tool toward other objectives. Email lists and websites, for example, alert constituents of collective actions, such as organized protests or letter writing campaigns.

Moving away from attention to social issues as confined to the realm of the individual, this approach builds on a model of social justice that assumes that differences in access to resources means that individuals do not share equal power to engage in political critique and action. Political-economic theories of media and of society guide this approach.

More than other approaches, media advocacy might be engaged by local community groups as well as by more resourced, larger non-profit organizations. There is potential for community involvement in working toward advocacy interventions, as well as mobilizing participants to act collectively toward media coverage and structural change. The source of information guiding the communication strategy is more likely to be based in community efforts than in a centralized agency. Typically having fewer resources than well-funded social marketing projects, media advocacy projects are more likely to attempt to attract news coverage, rather than purchase time or space.

Projects designate sources and activities designed to gain both media attention as well as sympathetic framing in order to confer legitimacy upon the perspective invoked. The assumption is that by setting the media agenda, public and policy agendas will follow (Gamson 1988; Olien et al. 1989). One goal may be to shift unquestioned assumptions in coverage of an issue to contested terrain; then groups can attempt to contest dominant visions of problems and communities. For example, for many years, the use of Arab characters primarily as villains was rarely contested in public discourse, but the American-Arab Anti-Discrimination Committee and other organizations have actively protested this characterization, moving the issue from uncontested to contested terrain (Wilkins and Downing 2002).

Once attention to an issue has been achieved, media advocacy attempts to shift coverage from being more episodic, focusing on personal stories, to more thematic frames, positioning the narrative within broader structural conditions. While the use of compelling anecdotes may be a useful way to gain attention, working toward structural change means promoting frames that highlight the broader conditions within which individuals act and social change might occur.

Interrogating Strategies for Social Change

Communication for social change can approach potential solutions through a variety of goals, addressing individual behavioral responses as well as norms and structures. In his insightful review of HIV/AIDS communication strategies, Tufte (2005) differentiates behavior change communication, a dominant approach positing the problem in individuals' lack of information, from "advocacy communication," working on behalf of the rights of those directly and indirectly affected by HIV/AIDS, and "communication for social change," attempting to address underlying causes such as poverty. Acknowledging what Tufte (2005, p117) describes as "a lack of conceptual clarity," this chapter seeks to expand upon an articulation of the advocacy potential of communication. First, though, it is useful to consider how communication fits into a broader trajectory of the social change project.

In the course of addressing social problems, offering mediated communication strategies as an intervention is but one approach. For example, if the problem of hunger is conceived as one of resources rather than information, then a media project might not be an appropriate response (Hornik 1988). Enforcement and engineering offer alternative approaches as well, encouraging more coercive or technological solutions (Paisley 1989). When including media as a central component of larger intervention strategies, it is assumed that the particular media technology is accessible, that the message is visually and verbally comprehensible, and that the mediated text will have some influence through its reception. Beyond the confines of a given approach, media for social change projects assume that the particular social problems addressed may be resolved through mediated strategies. The use of media not only responds to the logic of the particular condition addressed, but also resonates with broader ideological assumptions about social change.

The individualist preoccupation of social marketing and, to some extent, entertainment education builds on an ideological approach that privileges the individual as the primary actor in social change. In a market framework, individuals make choices about smoking cigarettes, using condoms, and recycling bottles. Conversely, in a social justice framework, policies that restrict smoking to certain areas, subsidize the cost of condoms, or facilitate the collection of recycling containers are seen as critical factors conditioning social change. If a woman is assaulted while running in a park, do we attempt to convince individual women to avoid public parks at night? Initiate public funding to increase lighting and enforcement patrols? Attempt to change the normative climate in which gender stereotypes assimilate persons into aggressive and victim roles? The potential solutions are constrained by descriptions of the problems.

The metaphor of the marketplace, grounding social marketing, builds from democratic and capitalist ideologies in which individuals are assumed to have equal abilities to vote and purchase. When people do not participate in elections or do not consume marketed products, the marketplace itself is not questioned, but individuals' lack of interest or knowledge is. Alternatively, a social justice orientation attributes social problems to structural conditions that inhibit participation.

Seeing information as never politically neutral (Rakow 1989), but as grounded within the institutional structures with the ability to construct and project particular perspectives, issues of power emerge in the process of strategic communication for social change. First, we must consider who has the right to create these communication campaigns, thus contributing to public discourse, not just in theory but also in actual practice. The right to voice is connected with issues of access to information, whether through strategic communication or other potential sources of information. In addition, we must consider who might benefit, as well as who might get hurt, from the proposed and actual changes advocated by strategic interventions (including but not limited to the funding organizations).

The point of intervention is to work toward social benefit, as designated by the organizations and groups creating the project. Communication for social change interventions may be seen as paternalistic (Salmon 1989) when assuming that the funding institution is better able to determine what is in the best interests of individuals or groups targeted for change. This paternalism may differ in degree, however, depending on the extent to which communities are engaged in the process.

Some counter this potential for paternalism with discussions of participation (Huesca 2002). What is meant by participation differs markedly across these models. Within social marketing, participation might be marked through consumption. Entertainment education along with social marketing might invoke the use of focus groups and formative research as a form of participation toward improving the execution and utility of the campaign. Apart from seeing participation as a means toward a previously constructed goal, advocacy communication articulates the importance of community participation in determining the priorities and salience of issues, as well as appropriate means for addressing them. When media advocacy builds from community concerns, these projects can be seen as engaging participation that involves prioritizing problems as well as constructing solutions.

Funding becomes a central concern not only in terms of offering resources needed for projects but also with regard to the potential consequences to the types of messages and strategies engaged. Donor funding in development has become increasingly privatized, with corporate "partners" often included in the composition of funding for social marketing and entertainment education. However, media advocacy projects may be directed against corporate agencies as limiting the potential to work toward the public interest. The concern is that by including corporate donors in the funding of projects designed for social benefit, the subjects addressed through these media projects may avoid potentially more controversial issues. In the interests of attracting corporate sponsorship of television and radio programs, issues such as racial tolerance, environmental conservation, human rights, sexual rights, and violence, might be more difficult to address; and if addressed, messages may be more likely than not to focus on individual decisions and acts rather than place responsibility on policies. An individualist approach, grounding participation within the metaphor of a global market, combined with concerns about paternalist control over decisions and economic agendas set by corporate institutions, limits the process of social change. In contrast, the proposed approach considers how communication might be used to advocate for progressive interests.

Advocacy Communication for and about Social Justice

The proposed conceptual model builds on critical assessments of the potential utility as well as limitations of communication approaches for social change. In contrast to social marketing's focus on immediate behavior change, advocacy communication attempts to influence structural and normative change to benefit marginalized and oppressed communities. Bridging the value of attracting audiences through entertainment-education approaches with media advocacy's social justice orientation, advocacy communication incorporates attention to news with popular culture, with the additional recognition of interactive communication technologies in contributing to social movement strategies. This model is described in terms of its goals, its conceptualization of communication as contributing to strategic intervention, and its use of research to enhance this process.

First, the central goal of this framework is to advocate for normative and structural change that works toward public benefit. Normative change involves creating a climate conducive to shifting the specific attitudes relevant to the strategic intervention, as well as ideas about potential self and collective efficacy. Attitudes might need to change gradually in ways that support a change, as well as reduce constraints. Promoting health benefits for same-sex partners, for example, means not only supporting equitable health benefits, but also reducing prejudice against same-sex partners. Moreover, people need to believe that there is the potential to create change in existing policies in order to activate energy and resources in order to motivate participation in the struggle. Structural change means creating policies and decisions within organizations and governments that structure actions. At this level of power, changing the rules of the game is critical in order to create more equitable and just playing fields.

These goals serve a social justice orientation, conceptualizing communication as a dialogic process, facilitating praxis, combining thoughtful reflection with informed action (Freire 1983). Communication is not perceived as limited to a hierarchical diffusion of information, or within horizontal connections across communities, but instead as facilitating activist strategies. Communication, then, represents a social and political process of creating and asserting meaning in a particular historical context, in which access to resources affords some groups more power than others to assert their voices.

Much of the language typically used to describe communication flows—as top down to represent diffusion models or as bottom up to convey grassroots models—can be seen as pejorative in that it positions one group as higher and therefore superior. Instead, I propose using a discourse of advocacy to convey the idea that communication can represent not just collective agreement, but also political resistance, with dignity and not subservience. This resistance is important in a framework that works to assert the rights of those who are marginalized and oppressed through supporting processes that promote justice and equity.

Although advocacy communication shares a social justice orientation with media advocacy, the approaches are distinct in terms of their strategic uses of media. A central limitation to media advocacy, along with social movement studies and communication studies, is that the existing literature focuses on attempts to attract and shift news coverage. This focus on news coverage is limited in two ways. First, there is increasingly a blur between the genres of news/information and popular culture (Wilkins and Downing 2002). Second, both news and popular culture contribute to political socialization. Social movements may target news to attract public attention to their goals, but public opinions are formed and reinforced within broader cultural climates, including but not limited to media sources. News media as well as popular culture contribute to what we know and think about particular topics, and how we might conceive of potential solutions and our ability to activate them. Similar to the critiques of those in the education-entertainment approach, advocacy communication recognizes that both in content as well as in reception news and fictional genres are not easily distinguished, and thus should be integrated within a comprehensive communication strategy.

The work of Soul City in South Africa represents an important illustration of a communication project bridging the work of entertainment education with media advocacy. Using an integrated media strategy built on radio and television dramas, the goals of this program extend beyond merely individual attitude and behavior change or attracting news media attention, toward affecting policy change built on community activism in areas such as the sale of cigarettes to children, workers' rights, and violence against women (Usdin 1998; Usdin et al. 2000).

There are several ways in which communication can facilitate advocacy. Communication can be used to encourage recognition of problems, as well as of potential solutions. This recognition helps to build toward mobilization, logistically organized through communicative channels. Mobilization toward collective action needs to be understood and acknowledged through public discourse. Decision-makers need to be aware of collective action taken, as well as reasons for that action. Communities need to be aware of decisions made through communication as well. But even more than an interest in making decisions transparent, communication should serve as a site through which groups contest interpretations of problems and proposed solutions. While the public sphere model offers a way of proposing communication as a site for pluralist dialogue (Jacobson 2000), advocacy communication recognizes that differences in access to resources create spaces through which some groups have more power than others to assert their perspectives. Understanding communications as a hegemonic process, there is room to negotiate and advocate changing conditions for a public good.

These potential uses for communication within the advocacy process can be served by a variety of media formats. News can serve as both a target for message dissemination as well as a vehicle for mobilization, as can web postings. Entertainment media serve to facilitate a cultural climate through which ideas can be projected. Interactive media, such as blogs, and interpersonal discussion groups can foster dialogue and plans for action. New computer technologies using programs that encourage sound-bite communication, such as Twitter, may enable logistical communication support toward mobilization, but may be limited in terms of allowing for more thoughtful communicative dialogue, structuring more episodic than thematic explanations.

One distinction concerns whether strategic intervention attempts to attract attention from mainstream media industries or controls content through funding and producing messages. Media advocacy strategies focus primarily on how to get attention in news in order to reach a broader audience through a channel, assumed to be seen as legitimate. There are other ways that an advocacy group might attempt to control content, such as paying for an advertisement or subsidizing scripts in popular culture. Scripts could be written by agents either paid for directly by an advocacy group or indirectly influenced through award programs rewarding attention to particular issues (such as population by the Population Institute and designated drivers by the Harvard School for Public Health). Controlling content may enable more thematic explanations of problems as well as more controversial perspectives to be known.

Each communication vehicle serves different interests. Reaching a broad public may be easier to do with public culture, whereas reaching a more targeted group of activists distrustful of mainstream media might be better championed through interactive computer technologies. Reaching the elite might best be done through attracting news media coverage. Interactive media are particularly useful when people do not trust the mainstream media, whereas news media might be valuable when attempting to mobilize a group attending to this genre and believing in its credibility. This is where the role of new communication technologies comes into play in helping to mobilize participants and advocate for particular causes. Communication for social change needs to address this literature more seriously.

Evaluation of advocacy communication must be engaged differently than assessments of individual projects, as is often done in studies of social marketing and entertainment education. Similarly, even academic studies of social movement organizations highlight the strategies of particular organizations. By focusing on singular strategies, these evaluations neglect attention to the critical issue being addressed. In order to resolve social justice problems within their historical and situational contexts, research needs to assess a variety of strategies and contextual conditions over time in order to consider long-term, sustainable solutions.

References

Barker-Plummer, B. (1995) "News as a political resource: Media strategies and political identity in the U.S. women's movement, 1966–75," *Critical Studies in Mass Communication*, 12(3): 306–324.

Escobar, A. (1995) *Encountering Development: The Making and Remaking of the Third World*. Princeton, NJ: Princeton University Press.

Freire, P. (1983) *Pedagogy of the Oppressed* (M.B. Ramos, trans.). New York, NY: Continuum.

Gamson, W. (1988) "Political discourse and collective action," *International Social Movement Research*, 1: 219–244.

Gumucio-Dagron, A. and T. Tufte (eds) (2006) *Communication for Social Change. Anthology: Historical and Contemporary Readings*. South Orange, NJ: Communication for Social Change Consortium, 556–560.

Hamelink, C. (2008) "Conversations with my robot," James Halloran Memorial Lecture. IAMCR, http://www.nordicom.gu.se/common/publ_pdf/283_hamelinkNY.pdf, accessed 30 October 2009.

Hemer, O. and T. Tufte (eds) (2005) *Media and Glocal Change—Rethinking Communication for Development*. Göteborg, Sweden: University of Göteborg/Nordicom.

Hornik, R. (1988) *Development Communication, Information, Agriculture and Nutrition in the Third World*. New York, NY: Longman.

Huesca, R. (2001) "Conceptual contributions of new social movements to development communication research," *Communication Theory*, 11: 415–433.

Huesca R. (2002) "Participatory approaches to communication and development," in W. B. Gudykunst and B. Mody (eds) *Handbook of International and Intercultural Communication*. Thousand Oaks, CA: Sage Publications, 499–518.

Internet World Stats (2010) "Internet usage statistics: The internet big picture," www.internetworldstats.com, accessed 23 August 2010.

Jacobson, T. (2000) "Cultural hybridity and the public sphere," in K. Wilkins (ed) *Redeveloping Communication for Social Change: Theory, Practice and Power*. Boulder, CO: Rowman & Littlefield Publishers, Inc., 55–72.

Jacobson, T. and J. Servaes (eds) (1999) *Theoretical Approaches to Participatory Communication*. Cresskill, NJ: Hampton Press.

Kotler, P., N. Roberto, and N. Lee (2002) *Social Marketing: Improving the Quality of Life*, second edition. Thousand Oaks, CA: Sage Publications.

Lerner, D. (1958) *The Passing of Traditional Society*. Glencoe: Free Press.

Melkote, S. R. and H. L. Steeves (2001) *Communication for Development in the Third World*, second edition. New Delhi: Sage Publications.

Montgomery, K. (1990) "Promoting health through entertainment television," in C. Atkins and L. Wallack (eds) *Mass Communication and Public Health*. Newbury Park, CA: Sage Publications, 114–128.

Olien, C. N., P. J. Tichenor, and G. A. Donohue (1989) "Media coverage and social movements," in C. T. Salmon (ed) *Information Campaigns: Balancing Social Values and Social change*. Newbury Park, CA: Sage Publications, 139–163.

Paisley, W. (1989) "Public communication campaigns: The American experience," in R. E. Rice and C. K. Atkin (eds) *Public Communication Campaigns*, second edition. Newbury Park, CA: Sage Publications, 15–38.

Rakow, L. (1989) "Information and power: Toward a critical theory of information campaigns," in C. T. Salmon (ed) *Information Campaigns: Balancing Social Values and Social Change*. Newbury Park, CA: Sage Publications, 164–184.

Rice, R. E. and C. K. Atkin (2001) *Public Communication Campaigns*, third edition. Thousand Oaks, CA: Sage Publications.

Rodriguez, C. (2000) "Civil society and citizens' media: Peace architects for the new millennium," in K. Wilkins (ed) *Redeveloping Communication for Social Change: Theory, Practice and Power*. Boulder, CO: Rowman & Littlefield Publishers, Inc., 147–162.

Ryan, C. (1991) *Prime-Time Activism: Media Strategies for Grassroots Organizing*. Cambridge: South End Press.

Salmon, C. T. (1989) "Campaigns for social 'improvement': An overview of values, rationales, and impacts," in C. T. Salmon (ed) *Information Campaigns: Balancing Social Values and Social Change*. Newbury Park, CA: Sage Publications, 19–53.

Schiller, H. I. (1991) "Not yet the post-imperialist era," *Critical Studies in Mass Communication*, 8: 13–28.

Schramm, W. (1963) "Communication development and the development process," in L. Pye (ed) *Communications and Political Development*. Princeton, NJ: Princeton University Press, 30–57.

Shah, H. and K. Wilkins (2004) "Reconsidering geometries of development," *Perspectives on Global Development and Technology*, 3(4): 395–416.

Singhal, A. and E. M. Rogers (1999) *Entertainment-Education: A Communication Strategy for Social Change*. Mahway, NJ: Lawrence Erlbaum Associates.

Snyder, L. (2002) "Development communication campaigns," in W. B. Gudykunst and B. Mody (eds) *Handbook of International and Intercultural Communication*. Thousand Oaks, CA: Sage Publications, 457–478.

Solomon, D. S. (1989) "A social marketing perspective on communication campaigns," in R. E. Rice and C. K. Atkin (eds) *Public Communication Campaigns*, second edition. Newbury Park, CA: Sage Publications, 87–104.

Tufte, T. (2005) "Communication for what? How globalization and HIV/AIDS push the ComDev agenda," in O. Hemer and T. Tufte (eds) *Media and Glocal Change: Rethinking Communication for Development*. Göteborg, Sweden: Nordicom, 105–122.

United Nations (2010) *Millennium Development Goals: At a Glance*. UN Department of Public Information, April 2010.

Usdin, S. (1998) "Interview with Dr. Shereen Usdin," Communication Initiative, http://www.comminit.com/en/node/149526.

Usdin, S., N. Christofides, L. Malepe, and A. Maker (2000) "The value of advocacy in promoting social change: Implementing the new domestic violence act in South Africa," *Reproductive Health Matters*, 8(16): 55–65.

Wallack, L., K. Woodruff, L. Dorfman, and I. Diaz (1999) *News for a Change: An Advocate's Guide to Working with the Media*. Thousand Oaks, CA: Sage Publications.

Wilkins, K. (1999) "Development discourse on gender and communication in strategies for social change," *Journal of Communication*, 49(1): 44–64.

Wilkins, K. (ed) (2000) *Redeveloping Communication for Social Change: Theory, Practice and Power*. Boulder, CO: Rowman & Littlefield Publishers, Inc.

Wilkins, K. and J. Downing (2002) "Mediating terrorism: Text and protest in the interpretation of *The Siege*," *Critical Studies in Media Communication*, 19(4): 419–437.

Wilkins, K. and B. Mody (2001) "Reshaping development communication: Developing communication and communicating development," *Communication Theory Special Issue*, 11(4): 1–11.

Chapter 7

Communication, Development, and Social Change

Future Alternatives

Florencia Enghel

Notes on a Paradoxical Field of Theory and Practice[1]

Communication for development (Servaes 1999, 2002), development communication (Wilkins 2008), and communication for social change (Gumucio-Dagron and Tufte 2006) are different approaches to defining a field concerned with the role of communication and media in strategic efforts to overcome societal problems common to many. When definitions within the field make reference to communities' self-initiated efforts, they may overlap to a certain extent with studies of community media (Howley 2009), citizens' media (Rodríguez 2001), radical media (Downing 2001), alternative media (Atton 2001), and/or media ethnography (Ginsburg et al. 2002).

Communication for development and social change refers to intentional and strategically organized processes of face-to-face and/or mediated communication aimed at promoting dialogue and action to address inequality, injustice, and insecurity for the common good. This, however, is only one definition among others. Inasmuch as understandings of communication and media interventions relate to debates and power struggles regarding development and social change, connected discourses and practices are contested matters. Attention to an active and changing vocabulary (Williams 1985) is thus of critical relevance in order to understand the field's past and present.

Besides constituting an area of academic theorization, research, and teaching, communication for development and social change is both a professional practice and an institutional project. "Project" refers here to the institutional influences that seek to give shape to the study and practice of development communication. Because of its origin, linked to the wider history of Western development policies (Harding 1998), over the years the framework of communication for development and social change as a project has been structured primarily through the intervention of bilateral and multilateral agencies. Unlike other fields, such as, for example, education, which involve a dimension of public policy-making, communication for development and social change work has rarely led to the formulation and implementation of specific policies at the national level (for an early discussion, see Servaes 1996).

Practitioners Wendy Quarry and Ricardo Ramírez (2009) rightly point out that "The only sustainable aspect of Communication for Development as a field is its constant negotiation for survival." Their statement could be understood to mean that the resources required for the field to progress—institutional, political, and economic—are permanently at stake (Deane 2004). Alternatively, it could be interpreted as characterizing the field as intersectional. Intentional communication interventions can advocate participation as a tool to promote a predefined goal efficiently and effectively, or as a valuable end in itself regardless of the desired results; they can favor individual behavior modifications or social and economic structural changes; and they can endorse commercial profit for some at the expense of others or promote solidarity and equality. Thus, the choice of direction at each possible crossroads is crucial (Williams 1985).

Perhaps it is this intersectional quality that makes the field seemingly incoherent, the broad range of practice possibilities apparently inconsistent, and the theoretical efforts fuzzy. Problems may also have to do with a certain insistence on understanding communication as an intrinsically positive tool, rather than as a complex and unequally distributed social capability with ambiguous possibilities (Garnham 1999; Sassen 2008). While the causes for shortcomings within the field remain unclear, the field's shortcomings could provide insights in terms of where to look next in order to formulate a substantial agenda for research.

This chapter situates the field of communication for development and social change in the context of the discourse and practice of bilateral and multilateral proponents and of what appear to be ongoing transformations in the political economy and structure of the international aid system. Contextualizing these elements provides the basis for a subsequent exploration of for whom and for what communication for development and social change could be mobilized in order to face several aspects of the current global crisis (for disaggregation of aspects of the crisis, see Amnesty International 2009, 2012; Doctors Without Borders/Médecins Sans Frontières 2009; UN 2010, 2011).

At a time when media and information have become commodities crucial for the expansion of global capitalism (Schiller 2007) while the shape and strengths of a much-predicated global citizenship supposed to be in the making remain unclear (Fraser 2008; Sassen 2008) and inequality is mounting (Wilkinson and Pickett 2010), I expect this contribution to be relevant to the broader context of communication and media studies.

Institutionalized Communication for Development

Bilateral and multilateral donors have acted as the primary funders of communication for development and social change projects and programs for several years and, as such, as the spinners of a related web of consultancies, publications, media developments, assessments, and research. In this respect, aid is never united. Practitioners and academics must address the tension between the influence of funders on project designs, research agendas and knowledge outcomes, and the need for financing for work in the public interest. Civil society organizations (CSOs) must resolve the problem of how to engage in collaboration with donors without sacrificing resistance to unfair conditions or surrendering transformative visions of change.

This chapter briefly describes the discourse and practice of communication for development and social change within three outstanding multilateral organizations: the United Nations (UN), which has worked with communication for development since the mid-1960s; the World Bank, which has done it at least since 1999; and the Organisation for Economic Co-operation and Development (OECD), which has recently shown a visible interest in the matter, connecting communication for development and social change with current debates on aid effectiveness.[2] In the case of the UN, I highlight divergent conceptualizations among agencies and persistent implementation problems. For the World Bank, I describe the rapid vanishing of its Development Communications Division and the poor outcomes of its highest profile initiative ever. In the case of the OECD, which gathers all those countries with significant bilateral aid programs under the umbrella of its Development Assistance Committee, I call attention to the promotion of a tight link between economic rationality and communication's deployment as a tool for efficiency.

The UN system goes backwards: from "achieving the Millennium Development Goals" to "demonstrating impact and positioning institutionally"

Communication for development has featured on the agenda of the UN for more than four decades. From 1967 to 1975, civil servant Erskine Childers, as head of the Development Support Communications Service at the United Nations Development Program (UNDP), advocated the integration of communication as a tool for effective program delivery within the UN (da Costa 2007). In 1969, communicator Colin Fraser was appointed to set up and lead a Development Support Communication Unit within the Food and Agriculture Organization (FAO), which until at least 1987 designed and implemented a number of innovative strategies in dialogue with rural communities across the world (Balit 2010, 2012).

However, today only a few agencies in the UN system work, to varying degrees, with communication for development. FAO has an excellent and well-documented track record of making use of communication for development in rural and agricultural settings to promote participation and improve living conditions, but concern that it is in the process of disinvesting in this area has been recurrent since the first World Congress on Communication for Development (WCCD) in 2006.[3] The United Nations Children's Fund (UNICEF), once the main exponent of communication for development in the UN, has prioritized instead high-profile external relations media initiatives in recent years (da Costa 2007).[4] Similarly, since 1999, and especially until 2005, "UNDP's Communication Office became an external relations machine, working with celebrities to raise the agency's profile and relegating development communication to the sidelines" (da Costa 2007). In 2006, the Oslo Governance Centre, a unit of UNDP's Democratic Governance Group, published a *Practical Guidance Note* with input from the Communication for Social Change Consortium[5] introducing a *communication for empowerment* framework (UNDP 2006). The approach was pilot-tested in a three-year initiative completed in 2010, but no further activities have been initiated since then.[6]

FAO, UNICEF and UNDP show different trajectories, understandings of, and approaches to communication for development. This is despite the fact that a mechanism exists to facilitate inter-agency exchange of communication for development ideas and experience and to promote cooperation: the UN Inter-Agency Round Table on Communication for Development, held on a biennial basis since 1988 (da Costa 2007). A review of documentation pertaining to the ninth, tenth, and eleventh round tables, held in 2004, 2007, and 2009, respectively (McCall 2009), reveals a series of persistent problems: a lack of resources for communication; confusion over how to understand and implement communication for development; strategies designed as an afterthought rather than as an integral part of planning and implementation processes; communication understood as public relations or dissemination; and a lack of coherence and coordination among agencies.[7]

In the wake of the WCCD, discussions during the tenth round table "focused on practical and achievable interventions around which UN agencies could develop a common approach, strategy and action plan for the implementation of C4D [communication for development] practice" (McCall 2009) to achieve the Millennium Development Goals (MDGs). However, by the time the eleventh round table took place in March 2009, jointly convened by UNDP and the World Bank, the MDGs were no longer on the agenda. Instead, the themes chosen—assessing and demonstrating the impact of communication for development, and institutionalizing communication for development in development agencies and organizations—echoed the WCCD's objective from three years before. The agenda was regressive. Moreover, unlike the WCCD, which was purportedly predicated on dialogue among different actors, the round table left civil society out of the discussions, held privately between 15 UN agencies and "key non-UN partners."[8]

The 2006 World Congress on Communication for Development: A Dead End

The WCCD, jointly organized by FAO,[9] the World Bank, and the Communication Initiative (CI),[10] was held in Rome on October 25–27, 2006 at FAO's headquarters.[11] Its purpose was "to effectively promote the mainstreaming of communication into development policies and programs and to promote strategic partnerships and alliances in this field." Its objective was "to demonstrate that communication for development is essential for meeting today's most pressing development challenges and, as such, should be more fully integrated in development policy and practice." The WCCD aimed to bring together communication for development practitioners, policy-makers, academics, global donors, non-governmental organizations (NGOs) and community representatives to share experiences and best practices in the field and develop "a common understanding of what works, what doesn't, and how best to measure impact."[12]

Preparations for the WCCD included issuing a call for papers and proposals, which were evaluated by a scientific committee.[13] Moreover, a series of regional consultations coordinated by FAO were carried out in Southeast Asia, the Near East, Sahel, Southern Africa, East Africa, Latin America, and Central America, purportedly "to obtain views and proposals of local practitioners and institutions for mainstreaming communication for development into sustainable development policies."[14] These consultations, led by local institutions, mobilized CSOs to a great extent.[15] Over 700 participants were invited to Rome for the event.[16]

In Rome, the contradictions between aims and reality were visible from the outset. Shortcomings were also soon apparent. Only a small fraction of the 137 papers selected by the scientific committee were fed into the event's program, and the outcomes of the regional consultations were not presented at the event's plenary sessions.[17] The output of the event, The Rome Consensus,[18] was the result of negotiations held behind the scenes among the three co-organizers rather than the product of much-talked-about collective participation and dialogue. Lacking specific goals and commitments, the consensus was clearly insufficient to serve as the basis for ulterior advocacy. Moreover, despite the expense and mobilization required to stage the WCCD, many of its recommendations turned out to be strikingly similar to those emanating from the 2004 UN Round Table (da Costa 2007).

In March 2007, the CI, in its *Final Report on the Survey on the WCCD*,[19] stated: "The WCCD is over and it is no longer the Secretariat or the Congress working groups who carry the responsibility for moving forward but all of us who believe that communication is essential to human development." With that statement, responsibility was placed everywhere and, thus, nowhere, revealing the WCCD as a rather inconsequential *mise-en-scène*. The WCCD website was discontinued,[20] and the information it once gathered is nowhere to be found. In the aftermath of the event, the World Bank's understanding of development communication remained instrumental, and its Development Communications (DevComm) Division, which had a prominent role in the organization of the WCCD and in 2008 was said to be "consolidating and strengthening its core of activities," has been scrapped from the organization's official website and may have been dismantled.[21]

Results Communication: The OECD Enters the Scene

The OECD's Development Assistance Committee (DAC) groups "the world's main donors, defining and monitoring global standards in key areas of development."[22] It comprises 24 member countries[23] with significant aid programs. The World Bank, the International Monetary Fund (IMF), and the UNDP participate in the committee as observers.[24]

The Informal Network of DAC Development Communicators (DevCom Network), created in 1988, brings together the directors of public affairs and communication of the foreign/development ministries and official agencies of the member nations of OECD's DAC, as well as new/emerging donors and multilateral institutions such as the European Commission, the UNDP, the World Bank, and regional development banks. "DevCom members ... shape thinking on how effectively to raise public awareness and support for development cooperation" through "the practice of communicating about development results."[25]

A working group on communicating results was assembled within the DevCom Network in 2007. The notion of *communication for results* was introduced in 2009. The conceptual framework put forward by the OECD likens communication for development to strategic communication, defining communication for results as "a tool as well as a process for the effective delivery of aid programmes" (da Costa 2009). Communication for results calls for the integration of communication *for* and *about* results as the way forward to institutionalize communication within the agenda derived from the Accra High Level Forum on Aid Effectiveness in 2008.[26] With the introduction of its framework for communicating results, the OECD seems to be taking the lead (formerly in the hands of the World Bank) in making visible efforts to instrumentalize communication in order to advance a specific agenda not necessarily concerned with justice, equality, or participation.[27] In parallel, the political economy and institutional structure of the international aid system are undergoing change.

A Changing Development and Aid System

Follow the Money

While bilateral and multilateral aid from official sources to so-called developing countries still represents a significant part of the picture (see, e.g., Easterly and Pfutze 2008), the privatization of aid seems to be on the increase. Data for 2008 from the SmartAid for Microfinance Index, an initiative of the Consultative Group to Assist the Poor (CGAP),[28] indicates that, while funding to support microfinance came primarily from bilateral and multilateral agencies (all of which were development banks), commitments from foundations increased by 73 percent "with the Bill and Melinda

Gates Foundation now the second-largest provider of grant financing overall."[29]

The Bill & Melinda Gates Foundation ranks as the most prominent private foundation in the US, both in terms of the size of its assets and of "total giving," at the time of writing.[30] According to its own records, in 2008 it distributed US$2.8 billion in grants and "direct charitable activities,"[31] with money going, for example, to the UN World Health Organization, the UN Capital Development Fund, and UNICEF. In 2009, money also went to the World Bank. As reported to the OECD, in 2009 it was "the third largest international donor to health after the United States and the Global Fund to Fight AIDS, Tuberculosis and Malaria."[32]

This indicates that the changes go beyond private aid being on the increase. On the one hand, we have the fact that in recent years multilateral agencies have established working relationships with the private sector (UN 2008) and received funding from private foundations. The reasons for this shift—namely, that funding for multilateral organizations is now coming from billionaires in the private sector rather than solely through governmental budget commitments—must be analyzed. On the other hand, we have the fact that the amount given by a private foundation can be high enough to make it one of the top decision-makers in a given area, such as, for example, health. The potential consequences of transferring to private hands the administration of matters of public concern, and thus the definition of social problems and their solutions, need to be anticipated (Sassen 2006; Wilkins 2008).

Meanwhile, debates over aid effectiveness remain fierce. In this context, the fact that bilateral and multilateral donors do not publish transparent data on how they spend aid money and the related operating costs despite their own calls for transparency is problematic for several reasons. On the one hand, the supposed beneficiaries of "the aid bureaucracies" (Easterly and Pfutze 2008)—that is, citizens of so-called low income or developing countries living in poverty—lack the information they would need in order to monitor and influence aid expenditure. On the other hand, while some CSOs try to push multilaterals to "publish what they fund,"[33] solid media coverage of development and aid matters is scarce or poor (Alfaro Moreno 2009), and studies of public attitudes toward aid in donor countries suggest an increasing concern with corruption and ineffectiveness (see, e.g., Glennie et al. 2012). There is a risk that, rather than getting involved in efforts to hold their governments accountable for the ways in which aid is allocated and used, citizens from bilateral donor countries will simply buy into the neoliberal mantra of philanthropy through "creative capitalism" proposed by billionaires such as Bill Gates, and be of the opinion that aid should be left to private riches to manage it as a business. The rise of private foundations adds further complexity to the scenario: while their work tends to have great visibility in the media, such foundations are legally and politically accountable to no one other than their own trustees.

Follow the Celebrities

In connection with these changes in the political economy and structure of the international aid system, researchers Richey and Ponte (2006; see also Richey and Ponte 2011; Richey 2012) highlight a reconfiguration of the modalities of international development assistance through their study of (RED), a trademark purportedly created to raise money for The Global Fund to Fight AIDS, TB and Malaria[34] through the sale of luxury goods. As documented by Richey and Ponte (2006), even if the best-known initiatives in public health development assistance are public–private partnerships, they are still financed mostly by traditional bilateral donors—and, to a lesser extent, by multilateral donors. However, (RED) gives individual consumers the impression that by spending money on luxury items they are single-handedly funding innovative ways to combat AIDS in Africa.

Through initiatives such as (RED), the problem of how to communicate development results—that is, how to educate citizens in donor countries about development and make the case that governmental aid to other countries is a relevant expenditure—could be obliterated. By pushing consumers to shop in order to "do good", celebrities such as Bono—the lead singer of U2 and (RED)'s iconic ambassador—are proposing a symbolic configuration in which the national state disappears as a relevant actor and development assistance is directly linked to the sales of transnational corporations. In the process, what should remain deep collective concerns are oversimplified and rendered consumerist individual matters.

Back to Communication: For Development or For Social Change?

As already noted, denominations for the field abound. Besides the academic definitions, of which three were referred to in the opening of this chapter, multinational and bilateral donors tend to introduce their own variations—ComDev (FAO), C4D (UNICEF), Communication for Empowerment (UNDP), DevComm (World Bank), and Communication for Results (OECD)—and we would find a similar situation if we looked into the denominations of choice among prominent NGOs.

While a few years ago it looked like "social change" was becoming the preferred denomination,[35] the shift from communication for development to communication for social change has not been linear (Wilkins 2009). In parallel, development remains a problematic concept (see, e.g., Sachs 1991; Nederveen Pieterse 2001 and 2009; Warah 2008, 2009). In this context, how do we name the field? Is there a choice to be made? When analyzed critically, the denominations "communication for development" and "communication for social change" show limitations. But vocabulary is not neutral, and attention to changes in the ways in which a key term is defined and to the ensuing arguments can illuminate the problems at stake within the field. From this perspective, choosing between development and social change would be unproductive. Following Williams (1985), I suggest that we allow a consideration of "the words as elements of the problems" to which they refer, and that we take into account the interconnections between differing or opposed concepts. The challenge of finding a better denomination as we try "to assert more comprehensive frameworks for strategic intervention" (Wilkins 2009) is not to be disregarded in acknowledgment of the relationship between how a problem is defined and which solutions are thus considered appropriate. In the meantime, referring to "communication for development and social change" will call attention to the fact that we are dealing with a contested field.

Mobilizing for Whom?

Overcoming the WCCD, or Mobilizing Beyond Donors' Agendas

The influence of bilateral and multilateral agencies in promoting communication for development and social change theorization and research that serves their agendas remains understudied. Likewise, the role of private philanthropists merits examination and critique. At the same time, the challenge of identifying areas for investigation beyond the limits posed by funders is pressing. Quarry and Ramírez (2009) refer to communication in development as a grey zone: "the sandbox where the conventional development industry allows us to play." The choice of words is provoking: the Oxford Dictionaries website describes a sandbox as a shallow or hollow box.[36] Their remark brings me back to the WCCD and, more broadly, to the role of funders in shaping knowledge and ways of knowing (Ginsburg and Gorostiaga 2003).

In advance of the WCCD, decision-makers in bilateral and multilateral aid agencies, national governments, and major NGOs were surveyed to identify the reasons for the slow progress in mainstreaming communication into development programs. The survey was a rerun of a 1994 study. Results showed that there was more confusion among decision-makers about communication for development—what it is and its possible uses—in 2006 than in 1994 (Fraser and Restrepo Estrada 2007; Fraser et al. 2007). Other studies argue that the notion of participation has long been abused (see, e.g., Thomas 2001) and that multilaterals resort to so-called participatory practice to sugar-coat projects that further impoverish the disenfranchised (see, e.g., Huesca 2002; Rush 2009).

But what about the role of scholars in this context? The above-mentioned scientific committee convened for the WCCD was tasked with writing a background study.[37] The study did, in fact, raise critical questions:

> Is there really light at the end of the long, dark tunnel of inequality? Why have we not yet resolved the key problems? Is there enough political will

> and commitment? Have any of the poverty alleviation solutions of the past really worked?
>
> World Bank (2007)

These questions, however, were not granted a plenary session in the WCCD's program, and therefore they were not shared or discussed widely. Dialogue between theorists, researchers, and decision-makers was not facilitated by the event's design. Moreover, in the aftermath the academics involved in the scientific committee failed to examine publicly the organizing conditions that constrained dialogue (Ginsburg and Gorostiaga 2003) and to make a statement in this regard. Mainly, the WCCD's failure to fulfill its commitments was either accepted uncritically or noted silently.

The forms of interaction between the development/aid conglomerate and academics in the field of communication for development and social change require critical analysis. The question of which actors in the field are currently exercising more power than others in terms of shaping knowledge and its dissemination needs to be addressed. In the context of a changing development and aid system, it is important that attention is paid to the origin of resources for research and the strings attached to funding.

If brought to the forefront, critical approaches that question the development industry's assumptions about "the nature of the problems addressed, appropriate solutions, and communities at risk" (Wilkins 2008) could illuminate the ways in which decisions about research topics, sites, methodologies, and findings (and their interpretation and social uses) are being made. Research needs to go beyond seeking to convince multilaterals that communication for development can be efficient in facilitating the delivery of this or that, or that participation can improve program effectiveness.

Overblown deference (Ginsburg and Gorostiaga 2003) to the needs of funders implies a risk of missing out on tackling serious problems that require attention. Moreover, researching to describe and evaluate projects or other forms of intervention is not enough. Mindful of differences, inequalities, and disconnects (García Canclini 2006; Hylland Eriksen 2007; Wilkinson and Picket 2010), research should focus on working to understand and change the conditions of increasing vulnerability and poverty that affect the lives of many citizens worldwide.

Mobilizing for the Different, the Unequal, and the Disconnected

Bilateral and multilateral agencies abstract their actions from the pressing material conditions in which they take place by referring to a standardized world composed of "beneficiaries" (emphasizing the supposedly positive outcomes of their interventions on individuals), "partners" (disregarding the unequal political and economic status of the recipient countries), and "representatives" (presenting CSOs as rightful brokers between citizens and foreign action). Such jargon obscures the role of citizens in struggles to improve the quality of their lives, and overrides a specific set of complex relationships among citizens, and between citizens and their national states, in the face of transnational intervention.

For anthropologist Néstor García Canclini (2006, p26), "communicational theories remind us that the connection and disconnect with others are part of our constitution as individual and collective subjects. Therefore, the *inter*-space, the space between, is decisive." From a cross-disciplinary, intercultural perspective,[38] García Canclini proposes a framework that in my view could give structure to a substantial research and advocacy agenda for communication for development and social change. I agree that the problem of "how to acknowledge the *differences*, right the *inequalities* and *connect* the majorities to the globalized networks" (García Canclini 2006, p14) is a theoretically relevant and politically crucial dilemma.

For García Canclini (2006, p14), in order to grasp the categories *different*, *unequal*, and *disconnected*, "it is necessary to consider how they complement each other and how they split." In his view, while anti-globalization movements have become strong enough to disturb neoliberal meetings such as Seattle (1999) or Cancún (2001), "rather than solve, they make evident the difficulties that persist when attempting to articulate *differences*, *inequalities*, *inclusion-exclusion* procedures and current forms of *exploitation*" (García Canclini 2006, p43). García Canclini (2006, p45) proposes that the task ahead of us is to figure out how to work jointly on these three processes, which in his view weave inter-culturality: the differences, the inequalities, and the disconnections.

The question relevant for me here is: How can communication for development and social change address this challenge? In García Canclini's (2006, p81) view:

> … it is difficult to imagine some sort of transformation towards a fairer regime without promoting policies that communicate those who are different (ethnic, gender, regional differences), righting the inequalities arising from those differences and from other non-equitable ways of distributing resources, and connecting societies with globally expanded information and cultural, health and wellbeing repositories.

This suggests the need for empirical research appropriately grounded in social theory that can, in turn, inform solid policy recommendations, as well as for detailed study of the relationship between intentional communication interventions funded by multilateral or bilateral development and the deployment of global capitalism as an economic project (Sassen 2006; McChesney 2007).

A Note on Globalization From Below and Global Citizenship

Referring to the dichotomy between nation states, major international organizations, and rich enterprises (globalization from above), on the one hand, and interpersonal relationships (globalization from below), on the other, anthropologist Thomas Hylland Eriksen (2007, p10) argues that "interpersonal 'globalization from below' is much more encompassing in shaping the world than often assumed." Hylland Eriksen highlights that while globalization is powered by economic and technological objectives, it takes place between people. He notes the role of trust in transnational web-based networks and calls attention to the fact that "people often use the opportunities offered by globalizing processes in unexpected ways" (Hylland Eriksen 2007, p14).

Interpersonal globalization from below and the ways in which communication is put to work strategically at this level in order to facilitate responses to economic globalization's wrongs remain understudied by communication for development and social change scholars. Disjunctures, discrepancies, fractures, contradictions, unexplained aspects, and unexpected outcomes deserve attention (García Canclini 2006; Hylland Eriksen 2007).

Future studies along these lines would best be undertaken while keeping certain considerations in mind. García Canclini (2006, p212) warns that "in a world where anonymous decisions made by the transnational market worsen the life conditions of the majorities, there is danger in the economic technocracy drowning political actors and reducing opportunities for subjects to exist to the imaginary scenes of the media."

More optimistically, Hylland Eriksen's (2007, p14) view is that "globalization creates a shared grammar for talking about differences and inequalities." While I agree with this observation, I find that talking about differences and inequalities is not enough. My view is that communication for development and social change should not settle for "reinserting those excluded." In designing new research agendas, we should be seeking to inform efforts to "transform unfair orders" (García Canclini 2006, p74).

García Canclini (2006, p214) highlights that we are still in the early stages of the process of understanding how a globalized citizenship could be. In his view:

> It is at this level of citizens' effective participation that the decisive importance of politics as the management of society reemerges—not as plain deals between hierarchies or media simulations of participation, nor as mere resistance by scattered actors or movements. In a world organized to both interconnect and exclude, the two politics most stage-managed so far for interculturality—tolerance of the different and solidarity from below—are requirements for coexisting. But if left at that, they risk becoming resources for coexisting with what we are not allowed to do. Communicating those who are different, righting the inequalities and democratizing access to intercultural legacies have become inseparable tasks for this time of miserable abundance.

Globalizers and the Globalized, Benefactors and Victims: Alternative Views

In defining for whom to mobilize, research in the field of communication for development and social change should broaden the selection, categorization, and definition of problems beyond those of donors. One way to do so would be to take into account the views of organized actors other than bilateral and multilateral agencies. For example, Amnesty International, a CSO that campaigns internationally for human rights, calls attention in its 2009 report *The State of the World's Human Rights* to the prospective human rights costs and consequences of the current economic crisis. In presenting its regional overviews, Amnesty International introduces "voice," a theme dear to communication for development and social change, as one of five categories under scrutiny together with deprivation, insecurity, exclusion, and accountability. More interesting in terms of opening up to topics beyond donors' agendas, in reporting by country the organization looks into whose rights are being affected most: those of children, indigenous peoples, lesbian, gay, bisexual and transgender people, migrants, women, and workers.

Another example is provided by Doctors Without Borders/Médecins Sans Frontières (MSF), an international medical humanitarian organization that delivers emergency aid to people affected by armed conflict, epidemics, healthcare exclusion, and natural or man-made disasters. In the 2009 edition of its annual list of the top ten humanitarian crises,[39] MSF proposes a mixed categorization, listing countries and problems. The purpose is to emphasize major obstacles to attempting to provide assistance to people affected by conflict in Pakistan, Somalia, Yemen, Sri Lanka, Afghanistan, Sudan, and the Democratic Republic of the Congo, and to call attention to stagnant levels of funding for treating HIV/AIDS as well as the ongoing neglect of other diseases such as Chagas, kala-azar, sleeping sickness, and Buruli ulcer, and the failure of the international community to combat childhood malnutrition,[40] which is "the underlying cause of up to half of the annual 10 million preventable deaths of children under five each year."[41]

Yet another example is given by La Vía Campesina, an international peasants' movement created in 1993.[42] Autonomous, pluralist, multicultural, and independent of any political, economic, or other type of affiliation, the movement unites 148 member organizations from 69 countries from Asia, Africa, Europe, and the Americas. Its main goals are to develop solidarity and unity among small farmer organizations in order to promote gender parity and social justice in fair economic relations; the preservation of land, water, seeds, and other natural resources; food sovereignty; and sustainable agricultural production based on small- and medium-sized producers.[43] The "who" of Vía Campesina are peasants, small- and medium-sized producers, landless people, rural women, indigenous peoples, youth, and farmers and agricultural workers. The "what" or main issues are agrarian reform, biodiversity and genetic resources, food sovereignty and trade, women's rights, human rights, sustainable peasant agriculture, and youth education.

These examples show that there are, in fact, informed, alternative ways of viewing social problems that could be incorporated when designing a research agenda.

Where to Look

Who are the protagonists of interpersonal globalization from below (Hylland Eriksen 2007)? Who are the different, the unequal, and the disconnected (García Canclini 2006)? Who are suffering the most in terms of lack of material resources, human rights, and dignity as welfare states are dismantled and nation states are "denationalized" in the process of privatizing what was once public (Sassen 2006; Wilkins 2008)? It is these subjects that a new agenda for communication for development and social change should be addressing, irrespective of the prescriptive positions of bilateral and multilateral donors. In expanding its scope, a revised agenda would, moreover, shift from a narrow focus on projects to concentrating on social problems (Wilkins 2009). Examples of subjects understudied by communication for development and social change include, but are not limited to:

- *Refugees, asylum-seekers, and internally displaced people*: according to the UN Refugee Agency, UNHCR,[44] at the end of 2008 there were 42 million forcibly displaced people worldwide, including 15.2 million refugees, 827,000 asylum seekers with cases pending, and 26 million internally displaced persons (UNHCR 2008).
- *Slum-dwellers*: since 2007, for the first time in human history, there have been more urban than rural inhabitants in the world. Most of the urban growth has taken place in poor countries (Hylland Eriksen 2007). Today, more than 1 billion people live in slums; in other words, "one in every three city-dwellers lives in inadequate housing with little or no basic services, and with the daily threat of insecurity, violence and forced evictions. Sixty per cent of the population of Nairobi, Kenya, lives in slums—one million of them in Kibera, the largest slum in Africa" (Amnesty International 2009).
- *Those without meaningful access to justice*: while international financial institutions have funded legal reform in the commercial sectors of several developing countries in order to strengthen the market economy, "there has been no comparable effort to ensure that poor people are able to assert their rights and seek redress through the courts for the violations committed by governments or companies." Amnesty International (2009) notes that, according to the UN Commission on Legal Empowerment of the Poor,[45] two-thirds of the world's population has no meaningful access to justice. In 2008, the commission argued that "four billion people around the world are robbed of the chance to better their lives and climb out of poverty because they are excluded from the rule of law" (Commission on Legal Empowerment of the Poor and UNDP 2008).

How to Look For What to Study

Literature on globalization tends to take the global/local duality for granted, thus failing to identify relevant dimensions of a complex phenomenon

(Sassen 2006). Within communication scholarship, an emphasis on media environments, on the one hand, and on the ubiquity of information and communication technologies (ICTs), on the other, has distracted attention from the particular social, political, and economic conditions in which people live.

Hylland Eriksen (2007 p. 141) notes that "globalization is always *glocal* in the sense that human lives take place in particular locations—even if they are transnational, on the move, dislocated." Such a premise calls attention to the importance of mapping and comprehending the contexts in which intentional communicative action can promote social transformation. In addition, it suggests that the specific experiences of certain groups could be explored in order to gain insights about particular approaches to tackling problems. Examples of such include, but are not limited to, migrants and youth.

The experience of certain migrant populations in terms of "establishing webs of security and trust independently of the state in which they live" (Hylland Eriksen 2007, p95), maintaining connections with relatives in their countries of origin, establishing connections with other migrants at the transnational level (interpersonal and based on personal commitment), as well as negotiating borders, is distinctive in terms of communication. In which ways are migrants communicating among themselves and with significant others at the local and transnational levels? What, for example, is the importance of remittances[46] to relatives at home? Which are the uses of cell phones, said to be spreading rapidly in the poorest countries of the world (Hylland Eriksen 2007; Ekström et al. 2012)? What does this tell us about the nature of citizenship under global conditions?

The experience of youth, characterized by extreme disparities in terms of resources across regions, countries, cities, neighborhoods, and population groups, with a large majority of the world's young people living in developing countries subject to inadequate education, unemployment, or insecurity in the labor market, is also distinctive in terms of communication (Tufte and Enghel 2009). In which ways are young people using communication and media technologies to organize locally and intervene in the public space? How do they bridge the gap between virtual and face-to-face communication? How are they voicing their feelings, views, and unmet needs to others (Enghel et al. 2009)?

Mobilizing For What?

In line with the already discussed variety of conceptual definitions, there is no agreement about what the core analytical dimensions of communication for development and social change are or should be. The pursuit of knowledge in the field tends to take place within the current dominant paradigm(s), while work beyond these borders remains rare (Shah 2007). A revised paradigm for the field could attract attention to what remains under-theorized and what has been neglected by research, and facilitate the conditions for identifying theoretical silences and debating disciplinary intersections

(García Canclini 2006). Mobilizing toward a substantially reconsidered paradigm would therefore appear to be a relevant direction forward.

Such a task is, of course, beyond the purpose and possibilities of this chapter (and would be best addressed collectively), but several of the elements discussed above could be structured toward a broader analytical framework for communication for development and social change. Premises to consider include:

- Research is political in its decisions about its directions and topics, the definition of the problems it chooses to study, the design and methods adopted, and the interpretation and utilization of the results of research (Ginsburg and Gorostiaga 2003; Wilkins 2008, 2009).
- In shaping the knowledge and the ways of knowing relevant to the field, analyses should not be reduced to sets of simplistic and/or generalizing conclusions (Ginsburg and Gorostiaga 2003; Hylland Eriksen 2007).
- Information and communication technologies—and their supposed potential to allow people to assert their rights and to see that they are, in fact, met—must be studied in their complex social contexts, acknowledging the immediate everyday life experience of people and communication's enmeshment with globalization as an economic project (Sassen 2006; Hylland Eriksen 2007).
- As stressed by Hylland Eriksen (2007), the fact that human lives unfold in specific locations is not overridden by globalization, and the continuous movement said to be characteristic of the phenomenon does not necessarily imply changes of a uniform speed or extent worldwide.
- The recognition of others, as well as meaningful social attachments, remains an underestimated and scarce resource that calls for an underscoring of the political and psychological importance of collective assembly (Hylland Eriksen 2007; McChesney 2007).

The Glocalization of Something, or Looking For the Future of Communication For Development and Social Change

As suggested in the chapter, to a certain extent the field's shortcomings could be the consequence of academic work increasingly trapped in the normativity of the development industry and out of touch with the variety of agents of change concerned with social injustice worldwide, who are deploying communication and media strategies to advance their struggles. As a matter of fact, my suggestion remains an empirical question that would merit research in itself.

Hylland Eriksen (2007) highlights the contrast established by sociologist George Ritzer between two contemporary tendencies. The *globalization of nothing* is defined by Ritzer (2004, p3) as "generally centrally conceived, controlled, and comparatively devoid of substantive content." Bilateral and multilateral communication for development, as discussed in this chapter, particularly in its most instrumental forms, falls into this tendency. The *glocalization of something*, in turn, refers to the ways in which a given "nothing" can be invested with something "in discriminating, critical ways" (Hylland Eriksen 2007, p59). In broadening its scope to avoid the "endogeneity trap" (Sassen 2006), communication for development and social change could advance collective solutions to pressing problems—facilitating critical conversations, channeling public concerns into decision-making bodies, reopening shattered dialogues, and articulating silenced voices. This would be one way to address the theoretical nothingness with which the field has been struggling in recent years.

Notes

1 An earlier version of this chapter was completed in October 2010, and most references throughout this version (finalized in December 2012) reflect information available at the time. Where it was possible to incorporate newer references without affecting the original structure of the chapter, I have done so. However, as a rule, because integrating newer data would have led to a different piece, I have refrained from attempting to update all references. For the earlier version, see Enghel (2011). Note the striking changes in most of the websites since first consulted in 2010: importantly, several missions and goals have been reworded, and whole sections have disappeared.
2 For a glimpse of aid effectiveness debates, see http://www.odi.org.uk/events/details.asp?id=495&title=serious-donors-about-aid-effectiveness-challenges-2010-beyond and http://www.stwr.org/aid-debt-development/reframing-the-aid-debate.html. The Swedish International Development Cooperation Agency has recently published a guidance note that directly connects communication to aid effectiveness (Sida, 2010).
3 At present, the role of the FAO's Communication for Development Team is to provide advice and technical assistance to member countries. See http://www.fao.org/oek/communication-for-development/field-programme/en/ for additional details and its definition of communication for development.
4 See http://www.unicef.org/cbsc/index_42328.html for UNICEF's mission and definition of communication for development.
5 A spin-off of the Rockefeller Foundation's work with communication for development and social change established as a US-based non-profit organization in 2003. See http://www.communicationforsocialchange.org/. For a critique of the Rockefeller Foundation as a funder of communication research see McChesney (2007, p29).
6 See http://www.undp.org/content/undp/en/home/ourwork/democraticgovernance/focus_areas/topics_civic_engagement/communication-for-empowerment.html and http://www.communicationforsocialchange.org/pdfs/c4e-globalreport-june2010.pdf.
7 The eleventh roundtable took place in India in 2011, hosted by UNICEF.
8 See http://www.undp.org/oslocentre/overview/ogc_communication_development.html.
9 Specifically, FAO's Communication for Development Group, at the time located in the Research and Extension Unit within the Natural Resources and Environment Department.
10 "CI is an extensive global network of people engaged in media and communication for development." See http://www.comminit.com/global/content/about-us-ci-global.
11 The idea was an outcome of the ninth United Nations Inter-Agency Round Table on Communication for Development organized in 2004 by the FAO with

UNESCO, the Government of Italy, the World Bank, the International Development Research Centre (IDRC), and the Centre for Agricultural and Rural Cooperation (CTA).

12 All direct quotes in this paragraph were retrieved from http://www.fao.org/sd/dim_kn1/kn1_060601_en.htm.

13 According to the World Bank (2007), the Scientific Committee reviewed 559 abstracts from 88 countries, subsequently inviting 290 authors to submit full papers, and 213 papers were submitted from 43 countries, of which 137 were accepted. No information is available regarding which countries were left out in the selection process.

14 See http://www.fao.org/sd/dim_kn1/kn1_060601_en.htm.

15 Based on such consultations, a *Compendium on Regional Perspectives* was prepared by Wendy Quarry and Ricardo Ramírez in October 2006. See http://www.comminit.com/es/node/303572.

16 According to the CI. See http://www.comminit.com/en/node/223102/306.

17 A respondent to the post-WCCD survey of participants set up by the CI stated: "a lot of work went into the regional consultations and it would have been useful to present them more fully at the initial plenaries … this was not adequately shared or brought fully into the actual WCCD discussions." Appendix to *The Communication Initiative Survey on the World Congress on Communication for Development* (p27), http://www.comminit.com/pdf/WCCD_Comments_Final2.pdf, accessed 30 August 2010.

18 See http://siteresources.worldbank.org/EXTDEVCOMMENG/Resources/Rome-Consensus07.pdf.

19 The *Communication Initiative Survey on the World Congress on Communication for Development* was conducted between December 2006 and January 2007 "to gauge responses" to the event. See http://www.comminit.com/global/content/world-congress-communication-development-lessons-challenges-and-way-forward.

20 The WCCD website could be found at www.devcomm-congress.org between December 2005 and January 2008. At present, it is no longer available at that address. It could also be found at www.devcomm.org between May 2006 and January 2008. For a partial archive, enter the web addresses into the Wayback Machine at http://archive.org/web/web.php.

21 See World Bank (2008, p8) for a definition of development communication. See World Bank (2008, ppxviii, xiv, 28) for a description of the DevComm Division.

22 See http://www.oecd.org/dac/, accessed 6 December 2012. In 2010, a former version of the website described the DAC as created for donor governments and multilateral organizations "to come together to help partner countries reduce poverty and achieve the Millennium Development Goals." Current wording indicates that "The DAC has played a role in forging major international development commitments, including the Millennium Development Goals." The reference to achieving these goals has been scrapped.

23 See http://www.oecd.org/dac/dacmembersdatesofmembershipandwebsites.htm.

24 See http://www.oecd.org/dataoecd/43/32/40986871.pdf.

25 See http://www.oecd.org/dev/communicationanddevelopment/aboutdevcom.htm and http://www.oecd.org/document/13/0,3343,en_2649_34101_39307533_1_1_1_1,00.html.

26 See http://www.accrahlf.net/.

27 For a critique of the OECD's role in the functioning of global capitalism, see Boltanski and Chiapello (2007).

28 "An independent policy and research center dedicated to advancing financial access for the world's poor. It is supported by over 30 development agencies and private foundations who share a common mission to alleviate poverty. Housed at the World Bank." See http://www.cgap.org/p/site/c/aboutus/ for more details.

29 This direct quote was taken from http://www.cgap.org/p/site/c/template.rc/1.11.45737/1.26.12206/, which is no longer available following a re-launch of CGAP's website. See http://www.syminvest.com/news/microfinance-funding-continued-to-grow-in-2008/2009/10/29/2130 for a secondary reference.
30 See http://foundationcenter.org/findfunders/topfunders/top100assets.html.
31 See http://www.gatesfoundation.org/annualreport/2008/Pages/overview-financial-statements.aspx.
32 See http://www.oecd.org/dac/47539494.pdf.
33 See, for example, http://www.publishwhatyoufund.org/.
34 "An international financing institution that invests the world's money to save lives"; see http://www.theglobalfund.org/en/.
35 The term "communication for social change" has been used by both academics (e.g., Tufte cited in Gumucio-Dagron and Tufte 2006; Servaes 2008) and non-governmental organizations (e.g., CFSC).
36 See http://oxforddictionaries.com/. Shallow: 1. of little depth; 2. not showing, requiring, or capable of serious thought. Hollow: 1. having a hole or empty space inside. 2. concave. 3. (of a sound) echoing. 4. lacking significance or sincerity.
37 Entitled *Communication for Development: Making a Difference*, the study was published as Annex 3 of The World Bank's publication *World Congress on Communication for Development*.
38 "Interculturality implies that those who are different are so in relationships of negotiation, conflict and reciprocal loans" (García Canclini 2006, p15).
39 See http://doctorswithoutborders.org/publications/topten/2009/.
40 As was evident at the World Summit on Food Security convened at FAO Headquarters in Rome in November 2009.
41 See http://doctorswithoutborders.org/publications/topten/2009/article.cfm?id=4127.
42 See http://www.viacampesina.org/.
43 In September 2010 La Vía Campesina publicly denounced the Bill & Melinda Gates Foundation Trust's recent acquisition of Monsanto shares (500,000 shares for over US$23 million), highlighting the fact that through its supposedly philanthropic projects the foundation is helping to open new markets for Monsanto, which is already the largest seed company in the world (Dena Hoff, North American coordinator of La Vía Campesina). See http://viacampesina.org/en/index.php?option=com_content&view=article&id=943:la-via-campesina-denounces-gates-foundation-purchase-of-monsanto-company-shares&catid=49:stop-transnational-corporations& Itemid=76.
44 See http://www.unhcr.org/4a375c426.html.
45 See http://www.undp.org/legalempowerment/clep_archive/index.html. The commission was in place between 2005 and 2008 and gave way to the UNDP's Initiative on Legal Empowerment of the Poor.
46 In 2003 "remittances from the US were significantly larger than the total value of official American development aid," and the total value of formal and informal remittances combined may exceed that of global foreign aid (Hylland Eriksen 2007, p80).

References

Alfaro Moreno, R. M. (ed) (2009) *¿Desarrollo? Encuentros y desencuentros entre medios y ciudadanía*. Perú: Observatorios en red/Calandria.

Amnesty International (2009) *Amnesty International Report 2009*. London: Amnesty International Publications.

Amnesty International (2012) *Amnesty International Report 2012*. London: Amnesty International Publications.

Atton, C. (2001) *Alternative Media*. London: Sage.

Balit, S. (2010) "Communicating with decision-makers," *Glocal Times*, 14 (May).

Balit, S. (2012) "Communication for development in good and difficult times," in F. Enghel, and K. Wilkins (eds) *Communication, Media and Development: Problems and Perspectives*, *Nordicom Review*, 33 (special issue). Gothemburg: Nordicom.

Boltanski, L. and E. Chiapello (2007) *The New Spirit of Capitalism*. New York, NY: Verso.

Childers, E. and M. Vajrathon (1968) "Development support communications for project support," in J. Servaes (ed) (2002) *Approaches to Development Communication*. Paris: UNESCO.

Commission on Legal Empowerment of the Poor and UNDP (2008) *Making the Law Work for Everyone (Volume 1)*. New York, NY: UN.

Communication Initiative (2007) *Final Report on Survey on the World Congress on Communication for Development*, http://www.comminit.com/pdf/WCCD_Survey Report_Final.pdf, accessed 19 April 2010.

da Costa, P. (2007) "Communicating as one? A look at the past and future of the UN's Communication for Development Round Tables," *Glocal Times*, 7 (February).

da Costa, P. (2009) *Study on Communicating Development Results*. Paris: OECD, www.oecd.org/dev/devcom/44836250.pdf, accessed 8 April 2013.

Deane, J. (2004) *The Context of Communication for Development, 2004*. Rome, Italy: FAO, Ninth United Nations Roundtable on Communication for Development, 6–9 September 2004, http://www.fao.org/sd/dim_kn1/docs/kn1_040701a10_en.doc.

Doctors Without Borders/Médecins Sans Frontières (2009) "Top ten humanitarian crises of 2009," http://www.doctorswithoutborders.org/publications/topten/2009/, accessed 3 September 2009.

Downing, J. (2001) *Radical Media: Rebellious Communication and Social Movements*. Thousand Oaks, CA: Sage Publications.

Easterly, W. and T. Pfutze (2008) "Where does the money go? Best and worst practices in foreign aid," *Journal of Economic Perspectives*, 22(2), Spring.

Ekström, Y., A. Hog Hansen, and H. Boothby (2012) "The globalization of the pavement: A Tanzanian case study," in F. Enghel and H. Wilkins (eds) *Communication, Media and Development: Problems and Perspectives*, *Nordicom Review*, 33 (special issue). Gothemburg: Nordicom.

Enghel, F. (2011) *Communication, Development, and Social Change: Future Alternatives*. Boston: International Communication Association Conference 2011.

Enghel, F., C. Flachsland, and V. Rosenberg (2009) "Youth, memory and justice: The Cromañón case and communication in an age of precariousness," in T. Tufte and F. Enghel (eds) *Youth Engaging with the World: Media, Communication and Social Change*. Gothenburg, Sweden: Nordicom.

Fraser, C. and S. Restrepo-Estrada (2007) "Policy-makers' perceptions of communication for development: Two surveys twelve years apart/personal reflections on what changed between 1994 and 2006 and its implications for advocacy," *Glocal Times*, 7 (February).

Fraser, C., S. Restrepo-Estrada, and L. Mazzei (2007) "What do they think? Policy-makers and the role of communication for development/a perception study conducted in 2005–6," *Glocal Times*, 7 (February).

Fraser, N. (2008) *Scales of Justice/Reimagining Political Space in a Globalized World*. Cambridge: Polity Press.

García Canclini, N. (2006) *Diferentes, desiguales y desconectados/Mapas de la interculturalidad*. Barcelona: Gedisa Editorial.

Garnham, N. (1999) "Amartya Sen's 'capabilities' approach to the evaluation of welfare: Its application to communications," in A. Calabrese (ed) *Communication, Citizenship, and Social Policy: Rethinking the Limits of the Welfare State*. Boston, MA: Rowman & Littlefield, 113–124.

Ginsburg, M. and J. Gorostiaga (eds) (2003) *Limitations and Possibilities of Dialogue among Researchers, Policy-Makers and Practitioners*. New York, NY: Routledge.

Ginsburg, F., L. Abu-Lughod, and B. Larkin (2002) *Media Worlds: Anthropology on New Terrain*. Berkeley, CA: University of California Press.

Glennie, A., W. Straw, and L. Wild (2012) *Understanding Public Attitudes to Aid and Development*. London: IPPR and ODI.

Gumucio-Dagron, A. (2009) "Playing with fire: Power, participation, and communication for development", *Development in Practice*, 19(4 & 5).

Gumucio-Dagron, A. and T. Tufte (eds) (2006) *Communication for Social Change Anthology: Historical and Contemporary Readings*. South Orange, NJ: Communication for Social Change Consortium.

Harding, S. (1998) *Is Science Multicultural? Postcolonialism, Feminisms and Epistemologies*. Bloomington and Indiana: Indiana University Press.

Howley, K. (2009) *Understanding Community Media*. London: Sage.

Huesca, R. (2002) "Tracing the history of participatory communication approaches to development: A critical appraisal," in J. Servaes (ed) *Approaches to Development Communication*. Paris: UNESCO.

Hylland Eriksen, T. (2007) *Globalization*. Oxford, UK: Berg Publishers.

Jacobson, T. and J. Servaes (eds) (1999) *Theoretical Approaches to Participatory Communication*. New York, NY: Hampton Press, Inc.

McCall, E. (2009) *Overview of UN Inter-Agency Round Tables on Communication for Development*. Washington, DC: UNDP and World Bank.

McChesney, R. (2007) *Communication Revolution/Critical Junctures and the Future of the Media*. New York, NY: The New Press.

Nederveen Pieterse, J. (2001; 2009) *Development Theory: Deconstructions/ Reconstructions*. London: Sage Publications Ltd.

Quarry, W. and R. Ramírez (2009) *Communication for Another Development: Listening Before Telling*. London: Zed Books.

Richey, L. (2012) "Mobilizing for global AIDS treatment," in F. Enghel and K. Wilkins (eds) *Communication, Media and Development: Problems and Perspectives*, *Nordicom Review*, 33 (special issue). Gothemburg: Nordicom.

Richey, L. and S. Ponte, (2006) "Better (RED)™ than dead: 'Brand aid', celebrities and the new frontier of development assistance," Danish Institute for International Studies (DIIS) Working Paper 26. Copenhagen: DIIS.

Richey, L. A. and S. Ponte (2011) *Brand Aid: Shopping Well to Save the World*. Minneapolis and London: University of Minnesota Press.

Richey, L., S. Ponte, and M. Baab (2009) "Bono's Product (RED) Initiative: Corporate social responsibility that solves the problems of 'distant others'," *Third World Quarterly*, 30(2).

Rodríguez, C. (2001) *Fissures in the Mediascape: An International Study of Citizens' Media*. Cresskill, NJ: Hampton Press.

Rush, J. (2009) "Sugar coating, or the manufacture of community support," *Glocal Times*, 13 (November).

Sachs, W. (1991) *The Development Dictionary: A Guide to Knowledge as Power*. London: Zed Books.

Sassen, S. (2006) *Territory, Authority, Rights: From Medieval to Global Assemblages*. Princeton, NJ: Princeton University Press.

Servaes, J. (1996) "Linking theoretical perspectives to policy," in J. Servaes, T. Jacobson, and S. White (eds) *Participatory Communication for Social Change*. New Delhi: Sage, 29–43.

Servaes, J. (1999) *Communication for Development: One World, Multiple Voices*. Cresskill, NJ: Hampton Press.

Servaes, J. (ed) (2002) *Approaches to Development Communication*. Paris: UNESCO.

Servaes, J. (2007) "Communication for development: Making a difference, a WCCD background study," in World Bank (ed) *World Congress on Communication for Development: Lessons, Challenges and the Way Forward*. Washington, DC: International Bank for Reconstruction and Development/World Bank.

Servaes, J. (ed) (2008) *Communication for Development and Social Change*. Los Angeles/London/New Delhi/Singapore: Sage.

Shah, H. (2007) "Meta-research of development communication studies, 1997–2005: Patterns and trends since 1958." Paper presented at the ICA Conference, San Francisco.

Sida (Swedish International Development Cooperation Agency) (2010) *Getting It Together*. Stockholm: Sida.

Sreberny-Mohammadi A. et al. (eds) (1997) *Media in Global Context: A Reader*. London: Bloomsbury Academic.

Thomas, P. (2001) "Freirean futures: Toward a further understanding of participatory communications," in M. Richards, P. Thomas, and N. Zaharom (eds) *Communication and Development: The Freirean Connection*. New York, NY: Hampton Press, Inc.

Tufte, T. and F. Enghel (2009) "Youth engaging with media and communication: Different, unequal and disconnected?," in T. Tufte and F. Enghel (eds) *Youth Engaging With the World: Media, Communication and Social Change*. Gothenburg: Nordicom.

UN (United Nations) (2008) *The United Nations and the Private Sector: A Framework for Collaboration*. New York, NY: United Nations Global Compact Office.

UN (2010) *Report of the Secretary General Keeping the Promise: A Forward-looking Review to Promote an Agreed Action Agenda to Achieve the Millennium Development Goals by 2015*, Document A764/665.

UNDP (United Nations Development Program) (2006) *Communication for Empowerment: Developing Media Strategies in Support of Vulnerable Groups/Practical Guidance Note*, http://www.undp.org/oslocentre/docs06/Communicationforempowermentfinal.pdf, accessed 10 November 2009.

UNDP (2011) *Human Development Report 2011*. New York, NY: UNDP.

UNHCR (United Nations High Commissioner for Refugees) (2008) *2008 Global Trends: Refugees, Asylum-seekers, Returnees, Internally Displaced and Stateless Persons*. Geneva: UNHCR.

von Feilitzen, C. (2006) "Enhancing media literacy," *Glocal Times*, 6 (December).

Warah, R. (ed) (2008) *Missionaries, Mercenaries and Misfits*. London: Author House.

Warah, R. (2009) "Development and other myths," *Europe's World*, http://www.europesworld.org/Default.aspx?TabId=809andPostID=843.

Wilkins, K. (2008) "Development communication," in W. Donsbach (ed) *The International Encyclopedia of Communication*. London: Wiley-Blackwell.
Wilkins, K. (2009) "What's in a name? Problematizing communication's shift from development to social change," *Glocal Times*, 13 (November).
Wilkinson, R. and K. Pickett (2010) *The Spirit Level/Why Equality is Better for Everyone*. London: Penguin Books.
Williams, R. (1985) *Keywords: A Vocabulary of Culture and Society*. New York, NY: Oxford University Press.
World Bank (2007) *World Congress on Communication for Development: Lessons, Challenges and the Way Forward*. Washington, DC: International Bank for Reconstruction and Development/World Bank.

Chapter 8

Beyond State-Centric Frameworks

Transversal Media and the Stateless in the Burmese Borderlands

Lisa Brooten

This chapter introduces the concept of "transversal media" created by stateless peoples to resist the ways in which sovereign state power defines and polices its own legitimacy through the erasure of their political voice and agency. Global communication scholarship would benefit from incorporating recent theoretical work in political science and border theory that challenges the state-centrism of much contemporary research on stateless peoples and offers new ways of thinking about human agency on a global scale.[1] Moving beyond existing global frameworks for media research, a focus on the transversal helps to illuminate those non-state, interstitial spaces that remain hidden in contemporary global communications research and in which much productive work is currently occurring.

To explore these ideas, this chapter will take as a case study the efforts of exiled Burmese journalists, refugees, migrant workers, and non-governmental organization (NGO) media examined in the context of larger, transversal dissident struggles.[2] Rapid changes in Burma since 2010 have resulted in the recent return of some exiled and dissident groups,[3] but this chapter focuses on developments in exile media since popular uprisings in Burma in 1988 that have arguably affected the direction of current changes in the country. It demonstrates how those defined as "stateless" and largely ignored or dismissed as passive victims often respond in frustration to the failures of corporate and state media to provide accurate representations of the complex political situations in which they find themselves. Over the last two decades, stateless activists and journalists from Burma have increasingly rejected the categorizations imposed on them by the sovereign power of both home and host states, and worked to remedy the images circulated about them on local, national, and international levels. These developments can help us to understand how those rhetorically and politically erased from the scene are working not only to become visible, but also to retool what is recognized as political under contemporary conditions of globalization.

Representations of the "Stateless"

> We are workers, migrants, refugees, Africans, East Europeans, Muslims, youths, single mums, queers, unemployed, lefties, anarchists ... We are ordinary people sick of being lied about. We are not "others" and we have the right to be here. We are here because you were there, plundering our countries and killing our planet, waging wars and arming dictators. The mainstream/corporate media have always sided with those in power, justifying their crimes and demonising their victims, softening up their brutality and blaming others. The result is that much-repeated lies become given facts and the victims of this screwed-up system are blamed for it.
>
> NoBorders Network (2010)

This quote was published on the NoBorders Network website, representing a European network of anti-border activists who resist migration control and demand freedom of movement for all people. It accompanied an announcement of a two-day protest organized in July 2010 against what they argue is a "racist press" that promotes hatred and fear of immigrants and others who are "different" by spreading rumors and lies. Their concerns were echoed at the time in the furor over Arizona's anti-immigrant law (SB1070), a potent example of the impact of inflammatory rhetoric about immigrant communities and their supposed threat to the state and nationalist identities worldwide.

While not as high profile a case, the situation facing the Burmese in exile has echoed many of these concerns. The image of the Burmese for many Thai people has been constructed from an early age through school history books focusing on ancient wars between Thailand and Burma. Yet, representations of the Burmese living in Thailand are negative in the contemporary Thai press as well, as several journalists and activists have argued. They cite media constructions of Burmese as criminals or simplistic hill tribe peoples (Sai Silp 2006) and, as one Thai journalist put it, that "the Burmese come to Thailand and kill Thai people" (Wandee Suntivutimetee, pers. comm., August 2, 2005).

Thai journalist Wandee Suntivutimetee conducted a study in which she found that in ten years of reporting on the Burmese in mainstream Thai newspapers, the representations were "mostly negative. [The Burmese] bring disease, cause crime, illegal migrant workers steal jobs from Thai people, they make stateless children" and so on (Wandee Suntivutimetee, pers. comm., August 2, 2005). In the week after a massive tsunami hit Thailand in December 2004, she found that of more than 400 articles, not a single one mentioned the community of Burmese migrant workers living in the affected areas, many of whom were killed or lost family members:

> Then after one week, the news that discusses Burmese workers is about the Burmese mobs who rob the Thais. After this coverage in *Khao Sod*

> [newspaper], the Burmese migrant workers cannot access relief goods. … So I think the case of the tsunami can show how the Thai media never cover [with] balanced news on Burma. They just produce prejudice.
>
> Wandee Suntivutimetee (pers. comm., August 2, 2005)

These media constructions and popular understandings of the Burmese living in Thailand reinforce discursive strategies that present stateless peoples as voiceless, passive, and helpless, on the one hand, or errant and deviant on the other (Soguk 1999; Nyers 2006). These tropes do very little to challenge sovereign state power to define who belongs, as they reinstate the inside/outside binary. They also do not examine the transversal struggles developing out of resistance by the stateless themselves, thus functioning to reinforce images of stateless peoples as both undeserving and incapable of political agency. This, of course, has tangible, practical implications for both the self-conceptions of the stateless and for how they are treated through official policy and in their daily interactions with people in the host society, such as state security forces, refugee and humanitarian workers, and local communities.

The State and the "State of Exception" in the Context of Modern Politics

While the study of states and their behavior is important, it has long been the primary focus in political science research, obstructing our understanding of how local politics is, in fact, increasingly global. The growing critique of this "methodological nationalism" recognizes that such an approach blinds researchers to the ways in which the power of the nation state has declined, to the fact that we live in a world increasingly defined by global-level risks, and that state-centric approaches to global politics have marginalized various forms of human agency, including and perhaps especially that of people defined as "stateless," rendering them invisible and un-theorized (Soguk and Whitehall, 1999; Blieker 2000; Beck 2004, 2007).

Dominant conceptions of modern politics remain spatial in nature, making crucial distinctions regarding insiders and outsiders, "between the citizens, nations and communities within and the enemies, others and absences without" (Walker 1995, p306). To secure the boundaries of a sovereign state requires violence, justified as a fundamental need in order to secure and protect the internal community from external threats. Thus, the sovereign state is paradoxical; to maintain peace requires violence, and efforts to secure internal harmony rely on a state of external chaos as their justification (Walker 1993).

The sovereign state has the ability to define both what is "normal" and what constitutes the "exception," and this power to conceptualize the "state of exception" is central to understanding how the sovereign state as well as the refugee and illegal immigrant are constituted (Nyers 2006, pxii). Yet the sovereign state itself is a social construction, a historical construct created and maintained through political activity, and recognizing this requires that we also see how national cultures are defined not only by their past, but by the continual reaffirmation of cultural identity. This is accomplished by continually reaffirming what constitutes the "state of exception" or that which is *not* the state (Nyers 2003, 2006; see also Carter 1994; Soguk 1999). With the production of the "refugee" and the "illegal immigrant" as non-citizens, sovereignty is produced and the state thus retains its legitimacy (Edkins 2000, p15).

In this way, refugees and "illegal" immigrants are not simply excluded from the political realm, but are rather trapped in a relationship of "inclusive exclusion," in which they are included in the definition of what is "normal" and part of the social "order" only through their exclusion from normal identities and membership in the sovereign state (Nyers 2006, pxiii). Deleuze and Guattari (1987, p386) demonstrate how this is a form of constitutive violence in that the state defines what constitutes its domestic political space through "a process of capture of flows of all kinds, populations, commodities or commerce, money or capital, etc." in which the state takes on the role of "town surveyor, converter, or highway interchange: the role of the engineer."[4]

There is the risk, however, that when stateless persons engage their political agency to define their own identity, the nature of sovereignty as arbitrary and contingent will be revealed (Nyers 2003, p2006). When those labeled "refugee" or "illegal immigrant" struggle with the sovereign hosting state in which they are identified as "Other," they help to illuminate how these categories are also social constructions. The state responds by working to confine the stateless to speechlessness and lack of political agency, and to allocate to them a "bare life with minimal requirements" so that they receive "only the basics of food, shelter and medical care to be sustained" (Nyers 2006, p98).

When the "state of exception" becomes the rule, what materializes is philosopher Giorgio Agamben's conception of "the camp," that space where the normal rule of law is suspended, and which houses those seen as barely human (Nyers 2006, p84). The non-state spaces of borderlands are arguably "camps," in which the "state of exception" becomes the norm, the "normal" rule of law does not apply, and inhabitants are conceptualized as less than human, a move that "takes on a political importance whenever we are confronted with situations where we are motivated to relate to each other as humans" (Nyers 2006, pp79–80). It is perhaps in this call to relate to each other as humans that transversal forms of communication and media make their greatest impact—by challenging common conceptions of the "Other" and sovereign state powers to exclude on this basis.

Beyond Methodological Nationalism and the Limits of Global Media Research

There has long been an unfortunate lack of exchange between the fields of political science and media studies (Downing 1996, 2008). While critical media studies have for the past few decades veered from the state-centric frameworks that remain the focus of mainstream political science research, recent critical political science scholarship has much to offer media theory. The limits of methodological nationalism have encouraged even mainstream political science scholars to theorize about the strains placed on state sovereignty by the processes of globalization, resulting in new political theories. Some of these move beyond the state-centric, *absolutist*, or "container" concept of space, in which societal space (the "nation" and its "culture") and geographic space (the "state") are seen to coincide (Sinclair and Cunningham 2000; Pries 2009).

Communication scholars began analyzing the role of powerful states in international communication and media in the 1970s with debates on cultural imperialism, and during the early 1990s in debates about the role of media as a driving force in globalization processes. More recent media scholarship has moved beyond the absolutist concept of states, making their relevance a topic of debate (Curran and Park 2000; Braman 2002; Atton 2003; Morris and Waisbord 2001; Morris and Meadows 2003). Transnational economic forces and transnational corporations have been the focus of critical political economy approaches to media, and other concepts such as post-national or postmodern forms of globalization and various forms of cosmopolitanism also address the limits of state-centric theories. Yet, the history of these theories and their associated practices is varied and controversial. The concept of cosmopolitanism, for example, as a playground of postmodern possibilities or "an inclusive, sophisticated and worldly demeanour" can be quite dangerous (Nyers 2003, p1073). As Nyers (2003, p1073) asks:

> Does the elevated status of cosmopolitanism—its narcissism, as it were—also rely on the construction of an abject other? ... All too often it is an "us"—Westerners, Europeans, humanitarians, etc.—who are the cosmopolitans, the champions of justice, human rights, and world order; leaving "them"—the Third Worlders, the global poor, the "wretched of the earth"—as the abject, the societies and subjects in crisis, the failed states in need of intervention.

Recent political science approaches have come to recognize the contributions of a *relativist* concept of social space that transcends borders and theorizes the emergence of new configurations which "extend above and between traditional container spaces of the concentric circles of local, regional, national, supra-national and global phenomena" (Pries 2009, p595). Yet, few of these political science and media theories shed light on the agency of stateless peoples and their media.

Perhaps most useful in this regard is the concept of diaspora, key in both political science and media studies to understanding how the relationship between the state and national identity is being challenged. The mass movement of peoples alongside the increasing circulation of media globally has led to the conceptualization of diasporic public spheres and their clearly transnational or even post-national consequences, as exemplified in Appadurai's (1996) notion of *mediascapes*, *ethnoscapes*, and *ideoscapes*. Media theorizing about diaspora has tended to focus on representations of ethnic and racial groups in mainstream media; yet research highlighting the agency of migrants themselves in media creation and consumption retains the state structure as a key framework as it details diasporic communities' negotiations between home and host states (Naficy 1993; Sinclair and Cunningham 2000).

Arguably, global media studies has led political science in the move from "a social problem or welfare conception of the migrant to an appreciation of cultural difference" and from "an essentialist or 'heritage' to a more dynamic, adaptive model of culture" (Sinclair and Cunningham 2000, pp13–14). Nevertheless, the term "diaspora" has been used in so many ways that it often glosses over significant differences such as class and gender (Clifford 1994; Sinclair and Cunningham 2000; Kraidy 2005). And as Sinclair and Cunningham (2000) point out, the diaspora is not coterminous with the concept of exile. Exiles, unlike diasporas, are not necessarily dispersed into multiple countries, and not all diasporas are caused by cataclysmic events, and while some in diaspora can return home, exiles cannot (Sinclair and Cunningham 2000, p19). Along this line, Appadurai (1996, p6) refers to "diasporas of terror" and "diasporas of despair" as distinguished from "diasporas of hope," but media scholarship has not yet contributed towards clearly illuminating these differences.

Missing from media studies approaches to understanding the complex contemporary world of global movement and mediation is the recognition emerging from critical political science scholars of transversal practices that emerge and flourish in border zones or interstitial, "non-state" spaces. Concepts such as "deformed cosmopolitanism" (Beck 2004) and "abject cosmopolitanism" highlight "the emerging political practices and enduring political problematics associated with refugee and immigrant groups resisting their targeted exclusion" (Nyers 2003, p1072–1073). These transversal processes are arguably components of international, transnational, and diasporic practices, but are unique and as yet unrecognized in media studies for their productive potential.

Transversal Practices in the Burmese Borderlands

In addition to maintaining cultural and affective ties with their homeland, the Burmese diaspora, much like other diasporic groups, engages in "an active social construction of new national symbols, practices, and identifications by political elites" (Adamson and Demetriou 2007, p510). This process of new identity construction would have been impossible inside Burma until very recently, as the generals of the country's military dictatorship (most recently, the State Peace and Development Council, or SPDC) strictly controlled discourse and manipulated the symbolic for their own political ends. And while the Burmese diaspora can be found in far-flung places, the country's borderlands have for two decades provided a vital environment for multi-ethnic efforts to construct a new vision of Burma.

It was to the borderlands that political dissidents fled after the military cracked down violently in response to a popular uprising in 1988 and again after the 1990 elections, which the opposition National League for Democracy (NLD) won in a landslide victory that the regime promptly ignored. It was in these borderlands that political dissidents from central Burma joined with the ethnic nationality groups, most of whom had struggled from the country's border regions against central government control since shortly after Burma's independence from the British in 1948. After the crackdown in 1988, the non-state, fluid, and unstable spaces along Burma's borders—under neither Burmese military nor neighboring country control—became the headquarters for the contemporary, multi-ethnic struggle against the Burmese regime.

Burma's ethnic nationality groups had long rejected the boundaries defined by the central government in their call for either a series of smaller, independent ethnically based states or a newly defined, multicultural federal system. This was a response to the state's violence in a process many claim has been a deliberate policy of ethnic annihilation or "Burmanization" of national identity. The multi-ethnic process of reconciliation in the borderlands since 1988 has led to a call for a federal system of semi-autonomous states as the best long-term solution to the country's history of ethnic conflict, a process of national identity reconstruction that could only occur in spaces beyond the state's control.

The consequences of the recent rapid changes in Burma remain to be seen, but the transversal practices of Burma's borderlands have forged the groundwork for a very different experience of political transition, for example, than that described by Downing (1996) in the countries of the former Soviet bloc, where there was no productive, pre-transition space in which to negotiate memory and history, ideology, tradition, and ethnic identity. The role of transversal media in memory and in the forging of a newly reconstructed national identity promises a very different outcome in Burma's case. It is in these border zones that people have forged new relations in response to their vulnerability in the face of the sovereign state powers that surround them, and have built the groundwork for a very different future than the one being shaped in sovereign state spaces. This is a process that *requires* the productive ground of "non-state," border zone spaces in which transversal processes flourish.

The Productive Potential of Liminal, Non-State Spaces

While the international trains our lenses on that which occurs between states, and the transnational focuses our attention on that which occurs above or beyond the control of states (while not necessarily rejecting nationalist identifications), the transversal focuses our attention on those non-state, liminal spaces between or beyond the control of sovereign state power.

These are the border zones, the non-places (but very material spaces) of the refugee camp, the detention center, and other fluid and unstable spaces where the stateless are (often literally) forced to hide from sovereign power. I make no claim that transversal practices are the most prevalent or most important contemporary political practices, just that they need attention if we are to recognize their productive potential thus far largely missed in the study of global communication and media.

Migrants and refugees are key players in transversal politics, participating in "'shared de-territorializations', zones of proximity, nonlocalisable spaces of Relation, that bring the underground and the aboveground, movement and home, inside and outside, into the proximity of one another to constitute complex spatial performances" (Soguk and Whitehall 1999, p680). These spaces and relations are established within what are often highly asymmetrical power relations but also provide opportunities for new relations, opportunities not available in the "Cartesian containments" of the sovereign state (Soguk and Whitehall 1999, p683). Rather than undermining or abandoning state forms of politics and international relations, transversal forms of political action instead reflect the transformation and redefinition of the state, including at times the rejection of its policed borders as well as its weakness or unwillingness to protect those within these borders.

Transversal practices fall outside the "container" spaces of sovereign states, and the physical spaces and the "non-state" nature of these border zones are key to their productive potential. Transversal practices ignore or reject borders, so their practitioners are vulnerable to the violence of those threatened by border transgressions; yet, it is arguably this vulnerability that provides a great deal of the productive potential of transversal practice. In other words, in rejecting or challenging established borders, transversal practices offer unique opportunities not found elsewhere—new forms of agency unavailable within the sovereign state—and this includes new forms of communication and media production and new forms of dissent.

Dissent and the Agency of the "Stateless"

Popular dissent is not new and has long been expressed through street protest, social movements, and other forms of civil disobedience. What is new is that such dissent is no longer primarily local in nature. The local intertwines with the global in new ways that give dissent a trans-territorial dimension and challenge the logic of national boundaries and "the spatial logic through which these boundaries have come to constitute and frame the conduct of international relations" (Keck and Sikkink 1998; Blieker 2000, pp1–2).

Yet, the theorization of dissent as applied to refugees and other stateless persons is limited. Nyers (2006) points, for example, to the category of "refugee warriors" found in refugee studies literature as a category defying traditional understandings of refugees as passive and voiceless. Refugee warriors, however, are defined by their acts of political violence, and Nyers

(2006, p100) notes that the limited conception of the relationship between the refugee and violence "is at once the limit of humanitarian discourses on refugees and also the limit of political discourses on sovereign spaces and identities."

Many Burmese activists in Thailand engage their political agency in non-violent ways, despite attempts by sovereign regimes of power to capture and control them. Burmese exiles in Thailand have consistently been threatened with repatriation and deportation, and have been victims of exploitation by local Thai authorities. While some dissident Burmese have qualified as "persons of concern" in Thailand, if they registered as such, they had to agree not to pursue political activities and must register to live in camps.[5] Others are eligible for resettlement in third countries through the United Nations High Commission for Refugees (UNHCR). Many others entered Thailand as migrant workers and could be arrested, detained, and deported at any time. Despite these risks, many chose to remain and continue with their political activities.

The stateless must work around the logic of the state and the constraints of local power to find the cracks through which to break from the "state of exception" and assert their agency and political subjectivity. In an important contribution to our understanding of the productive potential of transversal activism among border-based women activists from Burma, Mary O'Kane (2006, pp8–9) defines these women as "activist-refugees," a definition that fits many male activists as well:

> Specifically, their activism represents a particular process of deterritorialisation—that between activist *and* refugee. Activist-refugees straddle the ontological divide between citizen and non-citizen. They blur the barriers between the interlinking sovereign power regimes of the state and nation-state system. Yet, they cannot fit into the symbolic order of either and therefore carry the real potential to reveal the contingency and hence the insecurity of both narratives.

There are many examples of this. Next, I focus on several involving the media.

Media and Challenges to the State of Exception

Since the construction and maintenance of the state of exception requires the ability to create boundaries and categorize people as a means of inclusion and exclusion, a powerful means of resistance is therefore to refuse to recognize or work to counteract these categories upon which sovereign power relies (Edkins and Pin-Fat 2004). While media representations often act to create the very stereotypes that buttress these categories of inclusion/exclusion, they can also act as powerful means of combating such categories. In addition to challenging stereotypes, media can promote cross-cultural political acts of defiance against sovereignty's attempts to divide and separate.

The initial years of media development among the exiled Burmese, which began in earnest after 1988, saw three general trends: media developed as a means of international outreach, primarily to the US and Europe; media developed to improve communication within the Burmese opposition movement; and media that emerged to serve each of the country's major ethnic groups. After these efforts were well underway, journalists and activists recognized that their international outreach efforts had overlooked Burma's neighbors, who are key players in pressuring the Burmese military regime, but also as hosts to the Burmese in exile. In addition, while media emerged to serve the major ethnic nationality groups, work needed to be done to improve communication between the various groups constituting the political opposition to the Burmese regime.

As a result, the exile media began to make major efforts to translate information into as many languages as possible, including those of Burma's neighbors, especially those in Thailand. This began with the translation of key reports, such as human rights reports; but soon Burmese exile media began including translations of their news reports into Thai in addition to English and the various languages of Burma. The Shan Herald Agency for News (SHAN), for example, began publishing a four-part newsletter, with sections in Shan, Burmese, English, and Thai. They have recently begun publishing in Mandarin, as well. The Thai portion of the SHAN newsletter later split off into an independent publication, the *Salween Post*, founded by Wandee Suntivutimetee, the Thai journalist cited earlier who has studied Thai media representations of the Burmese, and who originally worked with SHAN. The *Salween Post* presents news and stories about Burma in Thai and works to deconstruct dominant stereotypes of the Burmese for Thai people. In another of many examples, the website of the Independent Mon News Agency (IMNA) provides news in Mon, Burmese, and English. As the borderlands have provided a space for the teaching of ethnic nationality languages until recently prohibited inside Burma, these media have been contributing to a reconstruction of Burmese identity as multicultural and multilingual, again in ways that would not be possible outside of the liminal spaces of the border. One important milestone to emerge out of the productive spaces of the border zones is the Burma News International (BNI) network.

Burma News International (BNI)

Burma News International (BNI) is a coalition of independent media groups founded in 2003 with the aim of "becoming a leading multimedia enterprise that presents a comprehensive picture of Burma and plays a role in promoting an understanding of the country." Its mission is to represent Burma "as a nation of diverse peoples" and help "to create knowledge and understanding about the different ethnic nationalities and regions of Burma" ("About BNI" 2008). With the exception of 2 of its 11 members (Mizzima and Network Media Group), all of the group's media partners are run by and targeted toward ethnic nationality groups.[6] Through its website (http://www.bnionline.net/), BNI operates as a central source for news from the rural areas of the country that are home to ethnic nationality groups. Its goals include increasing the amount and accuracy of reporting on ethnic areas of the country, opening up discussion between media groups, and supporting the development of media focusing on ethnic nationality communities. BNI supporters praise the network's policy of ethnic inclusiveness and decentralization.

BNI has played a key role in helping ethnic nationality media to raise their profiles through the network website, and in assisting its least developed members through funding and training opportunities. Some of this training has demonstrated the limitations of Western, state-centric approaches to "objective" journalism in the border zones, where ethnic nationality journalists must negotiate with ethnic nationality political leaders, Burmese military, and local host country officials (Brooten 2006). Nevertheless, in interviews, each member group reports benefiting from membership and acknowledges that perhaps the most exciting aspect of the network is its unparalleled potential as a collective. Working together, these 11 media groups offer the most comprehensive coverage of Burma of any Burmese media, given their unparalleled reach into the rural areas to bring to light stories that other media are unable to provide, including stories of the stateless in the border regions and neighboring host countries. BNI has functioned as a transversal network, a connection between many non-state spaces where political dissidence and "illegal" movement escape the control of any state.

The first collective project undertaken by the network was a nationwide survey conducted in 2008 in Burma's seven ethnic nationality states and six of the seven primarily Burman divisions (in total, 13 of the country's 14 administrative subdivisions), including in normally inaccessible ethnic minority areas. The survey assessed public attitudes prior to the national referendum on the military-backed constitution scheduled for May 2008. The survey was reportedly the most comprehensive and statistically representative poll of eligible voters ever conducted in the country, consisting of questionnaires and face-to-face interviews with over 2,000 Burmese voters, including farmers, traders, day workers, students, and housewives. The results showed a low level of awareness regarding the details of the military-drafted constitution, and widespread resistance to the coercive nature of both the state's constitution drafting process and the referendum procedure.

Then, on May 2, 2008, five days before a scheduled BNI press conference at the Foreign Correspondents Club of Thailand in Bangkok, Thailand, to announce the release of their survey, and eight days before the scheduled referendum, Cyclone Nargis slammed into Burma. This was the largest natural disaster to hit Burma in the country's recorded history, and despite the devastation, the military regime pushed on with the referendum, postponing it only in the hardest hit areas of the country. As widely expected, the military regime announced that more than 92 per cent of voters had approved the constitution. Unfortunately, the devastation of Cyclone Nargis overshadowed the importance of BNI's pioneering collective research effort. Nevertheless, the survey demonstrated the network's potential to cross borders and challenge the state's coercive practices by physically reaching out to report on people and conditions in hard-to-reach areas of this troubled country.

Since then, BNI member groups have continued to work collectively. In 2011, BNI released yet another report, this time on the 2010 elections (BNI 2011), and in 2012 yet another on the by-elections in order to "help voters tell the world about their expectations and experiences during the election process" (BNI 2012, p7). In addition, as the exile media have begun to return to the country in the wake of recent changes, BNI is working to provide support for those member groups who wish to set up offices inside the country. This network and its incredible potential would not have been possible without the productive transversal spaces of the border zones that have been home to many BNI member groups and reporters.

In addition to transcending the physical boundaries of the country's ethnic states, however, BNI has also functioned to reconfigure the boundaries of ethnic identity that have traditionally kept Burma's ethnic groups from trusting each other. The network works on this through the transversal practice of envisioning a multicultural Burma where all groups participate in a newly revisioned Burmese public sphere. This work would certainly not have been possible inside Burma, and likely would also have met with resistance in the heart of Thailand and other neighboring host states that must maintain official relations with the Burmese state. Operating in the liminal, non-state spaces of the borderlands, BNI members have worked to reframe the "us versus them" discourses that for so long have served the divide-and-rule strategies of the Burmese state.

This practice of recognition *across difference* can sometimes be found in transnational practices, but by no means always, as transnational practices often work to recreate and reinforce the boundaries of national identity in transnational spaces. Recognition across differences, however, *characterizes* transversal practice, and is arguably the very *core* of the transversal act.

MAP Foundation, Community Radio, and *Automatic Response Mechanism* (ARM)

> No human being is illegal.
>
> MAP Foundation motto[7]

Non-governmental organizations can also play key roles in transversal communication and media practices that challenge images of stateless peoples in the host state, as well as the state of exception that silences them and coerces them to refrain from enacting their agency. One example is the impressive work of the MAP Foundation in Chiang Mai, which focuses on providing services to migrant workers in Thailand and is staffed in large part by volunteers from the migrant community. A significant component of their work includes negotiating space for migrant and refugee agency within Thai communities and helping stateless peoples to function within and, at times, challenge the Thai system of rules and regulations for migrant workers by understanding and acting on their rights under Thai law. MAP also acts as a liaison between migrant workers, Thai employers, and Thai officials, attempting to solve problems and misunderstandings between different groups, and, in several instances, to take cases of employee abuse through the labor court and get compensation for workers. MAP is useful to Thai government authorities, who would find it very difficult to communicate in the languages required to register migrant workers without their help. In addition, MAP has had a great deal of support from people at the grassroots level of Thai society, as well as from the Thai National Human Rights Commission (Jackie Pollock, pers. comm., July 21, 2005).

MAP was founded informally in 1996 to address the obvious health and legal needs of migrant workers in Thailand. In a pattern established by several of the smaller NGOs working with the stateless Burmese, the initial years after its establishment saw the MAP Foundation working quietly under the radar, establishing credibility with the stateless by sharing with them the instability and threats of the non-state, liminal spaces in which they functioned. Although MAP's policy had always been to refuse to pay "tea money," for many NGOs, this involved befriending local Thai officials with financial or other material incentives, or in the case of media groups, providing information about the Burmese military unavailable but valuable to the Thais. Many of these groups eventually established themselves as legally registered organizations who could, in turn, challenge the often illegal practices of the local officials they were once forced to placate. Following this pattern, several years ago the MAP Foundation became a legal, registered NGO. As such, it plays a delicate balancing act by helping to negotiate between migrant workers and the local Thai officials who police Thai law and migrant worker policies. This evolution from under-the-radar to recognized NGO status is another example of the threats and instability that must be negotiated, but also the opportunities available to residents of the borderlands. It also attests to the fact that sovereign state power cannot be written off, but must be recognized and, at times, can even become a useful site of resistance, as when state sanctioned laws can be used to address problems facing the stateless (Tyler 2008).

Community radio stations in Thailand have offered opportunities for transversal media practices that challenge "us versus them" discourse and enable the stateless to enact their agency. In 1997, a team of Shan migrants at MAP in Chiang Mai, for example, began producing daily health education programs for the National Broadcasting Station of Thailand. When these programs proved popular, the station requested programs in Karen on health and culture. This provided a space to discuss issues of migrant health, such as the 16 part soap opera called *Love Never Dies* about living as a young migrant worker with HIV, produced by the MAP Karen radio team in 2000 (MAP Foundation 2006). While these programs were well received, they often aired at hours during which construction workers and other migrant workers could not listen. So using the material from the broadcasts and adding interviews and other information, MAP began producing audio magazines on cassettes and CDs, at first in Shan and Karen, and later in Burmese. When a group of migrants wrote a song about HIV/AIDS, MAP produced it; it was so popular that migrants asked for karaoke VCDs, which MAP then produced in Burmese, Shan, Karen, and Lahu (MAP Foundation 2006). MAP's radio programs and audio CDs were also played on several other community radio stations in Thailand.

MAP began broadcasting in 2004 on the community station FM99 in Chiang Mai, at first an hour a week, which eventually increased to four hours a week of live, bilingual programming in which information presented in Shan or Karen was then translated into Thai. This was intended to give Thai listeners a better understanding of what issues the Shan and Karen and other groups are facing (Jackie Pollock, pers. comm., July 21, 2005). The advantage of community radio is that programs are very local and can include live call-in shows where listeners can ask questions and offer opinions and advice. Thai employers often called in to ask or answer questions about procedures for registering workers, or to discuss the issues as they see them (MAP staff, pers. comm., July 21, 2005). Nevertheless, as the first Shan language community radio programmers in Thailand, MAP staff and volunteers were monitored very closely and had to avoid sensitive or political topics (MAP media coordinator, pers. comm., November 18, 2008).

In 2006, MAP staff stopped producing programs for the National Broadcasting Station of Thailand. While they recognized the value of these broadcasts, which could reach the entire country of Thailand as well as into central Burma and to Burma's border areas with Thailand, China, and India, to continue broadcasting these programs meant paying "tea money" to local Thai powerbrokers. MAP remained unwilling to do this, and it had also become a foundation with strict rules for financial transparency. As the MAP media coordinator told me, this decision was not because of problems with MAP's funders who may or may not have understood the need for "tea money," but "with our policy, because we are now a foundation" (MAP media coordinator, pers. comm., November 18, 2008). In refusing to play by the rules of these local powerbrokers, MAP staff and volunteers asserted a renewed sense of their collective identity.

In 2008, MAP took over FM99 from its previous Thai owner, and began broadcasting day-long programming that incorporated many more opportunities for members of the migrant community to participate and develop their own programs (MAP media coordinator, pers. comm., November 18, 2008). As MAP's media coordinator explained, the radio station is important for the migrant community because "they broadcast by themselves ... Inside Burma they cannot participate in broadcasting. But when they become migrants, it's a first [time] thing for them." Eventually, the migrant community has come to see the station as its own. As those who participate are fond of saying, "It's not MAP's radio program, it's the migrants' radio program" (MAP media coordinator, pers. comm., November 18, 2008).

Transversal practices respond to issues that very clearly cut across national boundaries, such as prejudice against migrants or violence against women. For example, one of MAP's projects has been a series of ongoing Women's Exchange (WE) meetings among stateless women of various ethnic groups. There was very consciously no action agenda in establishing these meetings, except to create a safe space in which women can discuss the difficult issues they are facing and to develop relationships across ethnic boundaries as well as confidence in their ability to express themselves publically. The monthly meetings began in Chiang Mai and the border town of Mae Sot, but then expanded to other places along the border. When a powerful tsunami hit Thailand in December 2004, decimating several migrant communities in the south, Women's Exchange groups were started among the Burmese migrant community there. One issue that the Women's Exchanges have worked to address in the last several years has been violence against migrant and refugee women, yet their work has also helped Thai women facing the same problems.

When two refugee women were raped outside of their camp by Thai camp guards, the Women's Exchanges began a discussion about how to deal with such problems. Making the case even more problematic was the fact that when the camp women's group took the rape victims to the local office of the UNHCR, they were told that the office was closed for the weekend (Jackie Pollock, pers. comm., July 21, 2005). Through discussions on what to do about cases of sexual violence such as this one, the women worked together with MAP to put together a book entitled *Automatic Response Mechanism* (ARM), which outlines ten steps migrant and refugee women can take if they have been sexually abused (Jackie Pollock, pers. comm., July 21, 2005). This includes steps as simple as speaking with a friend, as well as more complicated processes such as pressing charges and taking a case to court, and tries to prepare women for the difficulties of the process by outlining the pros and cons associated with each action. MAP co-founder Jackie Pollock (pers. comm., July 21, 2005) explained the three-year process involved in the book's creation by women of many different ethnic, linguistic, socio-economic, and educational backgrounds:

> What we did was put it together and then go back to the Women's Exchange groups, and say, "Does this work for you? ... Does it make sense to put this?" And then [we'd] change it, changing all the languages, [and then] go back to them. Then once we got it out like this, then last year ... we went up and down the border doing a five-day training on it. [We explained to women:] "This is what we can do, how we actually implement that." I mean, [in] those trainings, we took the women to the one-stop crisis center in their town so they could meet the setup people in the hospital.

At the time MAP was conducting this training with stateless women in Thailand, several of the hospitals they visited began setting up crisis centers for victims of sexual or violent abuse; these centers, once open, were then available for all women, including Thai women who previously had had no such service.

After the women participating in the Women's Exchange groups had painstakingly produced the first edition of ARM, they realized that the steps relevant for migrant women who must negotiate Thai law would not be relevant for those refugee women who lived in camps controlled by the UNHCR. About this time, in response to the situation in Thailand as well as international pressure to deal with issues of sexual violence, the UNHCR decided to put together their own guidelines for how to deal with sexual violence against refugees. Two UNHCR representatives approached MAP to ask for assistance, but their idea of the project did not match MAP's conception of how to work with women in the border areas. Jackie Pollock (pers. comm., July 21, 2005) describes their first meeting with UNHCR representatives:

> In Mae Sot, we ran into two very nice women. They asked us to go through the training for their protocol, of course, for the protocol in the camp, which was great. So we put a training program together, which would have been like so much of what we've done at the beginning of this [process]. And they said, "Oh, no, no, you can't do it for that long in the camp. No, no, no, all you need to do is tell them what the protocol is." So we [replied that] that's not how we work. You know, that's why it took three years to put this together, because it's the women trying to work it out themselves.

When I spoke with MAP staff in 2008, the UNHCR had agreed to work with MAP, the American Refugee Committee (ARC), and the Karen Women's Organization (KWO) in a process of community consultation to develop their own guidelines for how to respond to sexual violence against refugees, based on the model offered by ARM. However, by mid-2010, the women in the camp were reportedly using ARM instead of the more recently developed UNHCR guidelines, in large part because many of them had been involved in the process of creating ARM and felt ownership over the process, which they also understood better after having received training in how to use ARM (Jackie Pollock, pers. comm., June 17, 2010).

By addressing threats such as arrest, deportation, prejudice, and violence that cut across identity boundaries, the work of various individuals, media groups, and grassroots NGOs such as MAP challenges the state of exception imposed from above by sovereign powers in several ways. For one, many border-based groups, such as the group that produced ARM, have refused to create hierarchies of difference between ethnic nationalities, or between Thais and non-Thais, recognizing, for example, female victims of violence as equals and as similarly traumatized by the violence they face. This work also clarifies how the sovereign state and the very international organizations charged with protecting stateless peoples have not been effectively doing so, highlighting the ability of transversal practices and resistance to the state of exception to pinpoint the weakness or unwillingness of sovereign powers to protect people on the ground. And the transversal practices highlighted here and made possible only in the non-state, liminal spaces of the borderlands also work to create solidarity among stateless peoples and between them and the "legal" citizens of other states, establishing them as collaborators, co-researchers, or productive agents of their own lives rather than passive, voiceless victims. This is a direct challenge to the sovereign powers of the state and the state of exception. It is an embrace of transversal politics and the shared deterritorializations and "spaces of Relation" made possible in border zones, where the "countless non-heroic practices that make up the realm of everyday life," in fact, carry great power and potential for change (Blieker 2000, p278).

Concluding Thoughts

The changes ushered in by transversal dissent, however powerful, do not result in immediate change; rather, they work slowly and indirectly, gradually transforming societal values in a process that can often be difficult to see or prove, and which never ends (Blieker 2000). As Blieker (2000, p281) notes:

> There is no emancipatory peak to be climbed. Dissent is the very act of climbing, daily, doggedly, endlessly. It is not an event that happens once, a spectacular outburst of energy that overcomes the dark forces of oppression and lifts liberation into a superior state of perpetual triumph. ... Even the most just social order excludes what does not fit into its view of the world. Inclusiveness lies in a constant process of disturbing language and rethinking meaning, rather than in an utopian final stage.

Downing (2001) urges attention to the "slow burn" impact of dissident media, in contrast to the focus on the instantaneous that characterizes much media research. He argues that the energy put into and derived from media projects such as these continues on in other projects over decades in ways that cannot be measured yet should not be underestimated. The role of such media, he maintains, especially if the struggle seems prolonged, "is to keep alive the vision of what might be, for a time in history when it may actually be feasible" (Downing 2001, p9).

While state sovereignty remains the hegemonic form of political organization, transversal dissent by communities whose actions and identities are no longer primarily state centric but, rather, have shifted to cross identity boundaries is one of the most important developments for understanding how politics is being transformed today. Burmese media groups, political activists, migrants, and refugees, while challenging the state of Myanmar and, at times, the sovereign state power of host states, have not been openly challenging the state system itself. In fact, they generally desire to rebuild a community within such a sovereign state system (O'Kane 2006). Yet, the work of activists in exile offers insight into "the intrinsically co-constitutive relation between the 'informal' and 'formal' political spaces and how they transform each other" (O'Kane 2006, p11).

Scholars warn against re-inscribing through their work the very powers they are trying to fight; this occurs when scholars a-critically prioritize sovereignty and the actions of humanitarian agencies that reproduce sovereign political relations by reinforcing state-centric views of refugees as apolitical (Nyers 2003, 2006; Tyler 2006, 2008). Critics of this tendency call on us to transform the pejorative stereotypes of the stateless as speechless or invisible, or dangerous less-than-humans, and to include stateless peoples as central partners in decision-making regarding their lives (Nyers 2006; Tyler 2008). They call on us to recognize the power of their political agency and how its resistance "to modern sovereign subjectivities runs the risk of opening up the entire architecture of our received conception of the political" (Nyers 2006, pp129–130). We need to explore "how the citizen–human dynamic is being renegotiated by 'refugees,' irrespective if they are 'legitimate'" (Nyers 2006, p130), and the importance of non-state spaces for the emergence of transversal dissent, including transversal media. There is a strong sense of hope in this, and I join those who call on us to "be open to refugee politics, give support to it, and take direction from it" (Nyers 2006, p131).

Notes

1 The author would like to acknowledge the debt this work owes to the work of political scientist Mary O'Kane (2005, 2006).

2 The name "Burma" has taken on political connotations since 1989, when the military changed the country's name to the Union of Myanmar. While the government claimed that the new name is ethnically neutral and promotes national unity, the opposition rejected the name change, made without consulting the people. The opposition has generally preferred "Burma" and the terms have become to some degree an indication of political positions in the struggle for control over the country. Since this chapter focuses on the opposition movement, the term "Burma" will be used here. The term "Burman" is used to refer to the ethnic majority of the country, while "Burmese" is used as an adjective to refer to the language and the country's various peoples. Thus, an ethnic Karen, for example, is both Karen and one of the Burmese peoples.

3 Elections in November 2010, although highly problematic, brought a nominally civilian government to power. Incremental reforms since, including the release of political prisoners, resulted in a December 2011 visit by US Secretary of State Hillary Clinton, the highest-ranking US official to visit Burma in 50 years; the US then sent its first ambassador to Burma since 1990. The opposition National League for Democracy (NLD) won nearly all of the small number of parliamentary seats in the 2012 by-elections, including NLD leader and Nobel Peace Prize Laureate Aung San Suu Kyi, who nevertheless joined a Parliament still

overwhelmingly controlled by the military-backed ruling party. The US has also lifted a travel ban on senior military leaders and partially eased US sanctions, and President Obama visited Burma in November 2012, the first sitting US president ever to do so.

4 The hosting of refugees also involves violence. Derrida's writings on the topic of hospitality demonstrate how unlimited hospitality is an acceptance of risk, in that unlimited hospitality requires that one take in anyone or anything that desires entry, regardless of identification or even of human form. Therefore, the hospitality as practiced in hosting refugee populations is limited in form, and violent in that it is "always aware of another way of relating with others, an approach that does not insist on conditions—such as proper documentation, state citizenship, or … even 'humanity'" (Nyers 2006, p72). Derrida has introduced the concept of "hostipitality" to recognize the hostility inherent in the practice of hospitality, to recognize the way that limited forms of hospitality of necessity involve animosity, hostility, and antagonism (Nyers 2006, p73).

5 The Thai government does not officially recognize refugees on its territory as it was not a signatory to the United Nation's 1951 Convention on the Status of Refugees or its 1967 Protocol. Thus, the term "person of concern" is used by the Thai government and the UNHCR to refer to those displaced from Burma who, under the terms of the convention, qualify for refugee status.

6 The nine BNI members that cater to ethnic nationality groups are the *Kaladan Press* (Rohingya), *Khonumthong News* (Chin), *Narinjara News* (Arakan), *Kachin News Group*, *Independent Mon News Agency*, *Kao Wao News* (Mon), *Kantarawaddy Times* (Karenni), *Karen Information Center*, and *Shan Herald Agency for News*.

7 Registered under Thai law and known in Thai as "Foundation for the Health and Knowledge of Ethnic Labour," the MAP Foundation began as a small grassroots group called the Migrant Assistance Program, and later registered formally as a non-governmental organization, retaining its English acronym MAP, which was familiar to the local communities with whom they work.

References

"About BNI" (2008) http://bnionline.net/index.php?option=com_contentandtask=viewandid=1andItemid=38, accessed 17 September 2008.

Adamson, F. and M. Demetriou (2007) "Remapping the boundaries of 'State' and 'National Identity': Incorporating diasporas into IR theory," *European Journal of International Relations*, 13(4): 489–526.

Appadurai, A. (1996) *Modernity at Large: Cultural Dimensions of Globalization*. Minneapolis, MN: University of Minnesota Press.

Atton, Cs. (2003) "Infoshops in the shadow of the State," in N. Couldry and J. Curran (eds) *Contesting Media Power*. Lanham, MD: Rowman & Littlefield, 57–69.

Beck, U. (2004) "Cosmopolitical realism: On the distinction between cosmopolitanism in philosophy and the social sciences," *Global Networks*, 4(2): 131–156.

Beck, U. (2007) "The Cosmopolitan condition: Why methodological nationalism fails," *Theory, Culture and Society*, 24(7–8): 286–290.

Blieker, R. (2000) *Popular Dissent, Human Agency and Global Politics*. Cambridge: Cambridge University Press.

BNI (Burma News International) (2008) *Findings from Survey on the May 10 Referendum*, 7 May. Chiang Mai, Thailand: BNI.

BNI (2011) *Hobson's Choice: Burma's 2010 Elections*. Chiang Mai, Thailand: BNI.

BNI (2012) *By-Elections or Sell Elections: The Pre-Election Observations to Burma/Myanmar 2012 By-Elections*. Chiang Mai, Thailand: BNI.

Braman, S. (2002) "A pandemonic age: The future of international communication theory and research," in W. B. Gudykunst and B. Mody (eds) *Handbook of International and Intercultural Communication*, second edition. Thousand Oaks, CA, and London: Sage Publications, 399–413.

Brooten, L. (2006) "Political violence and journalism in a multi-ethnic State: A case study of Burma (Myanmar)," *Journal of Communication Inquiry*, 30(4): 354–373.

Carter, D. (1994) "The art of the state: Difference and other abstractions," *Journal of Historical Sociology*, 7(1): 73–102.

Clifford, J. (1994) "Diasporas," *Cultural Anthropology*, 9(3): 302–338.

Curran, I. and M. I. Park (2000) *De-Westernizing Media Studies*. London: Routledge.

Deleuze, G. and F. Guattari (1987) *A Thousand Plateaus: Capitalism and Schizophrenia*. Minneapolis, MN: University of Minnesota Press.

Downing, J. D. H. (1996) *Internationalizing Media Theory: Transition, Power, Culture*. Thousand Oaks, CA: Sage Publications.

Downing, J. D. H. (2001) *Radical Media: Rebellious Communication and Social Movements*. Thousand Oaks, CA: Sage Publications.

Downing, J. D. H. (2008) "Social movement theories and alternative media: An evaluation and critique," *Communication, Culture and Critique*, 1(1): 40–50.

Edkins, J. (2000) "Sovereign power, zones of indistinction, and the camp," *Alternatives*, 25(1): 3–25.

Edkins, J. and V. Pin-Fat (2004) "Introduction: Life, power, resistance," in J. Edkins, V. Pin-Fat, and M. Shapiro (eds) *Sovereign Lives: Power in Global Politics*. New York, NY: Routledge.

Keck, M. E. and K. Sikkink (1998) *Activists beyond Borders: Advocacy Networks in International Politics*. Ithaca, NY: Cornell University Press

Kraidy, M. (2005) *Hybridity, or the Cultural Logic of Globalization*. Philadelphia, PA: Temple University Press.

MAP Foundation (2006) *1996–2006: No Human Being is Illegal, No Migrant Worker is Illegal*. Chiang Mai: MAP Foundation.

Morris, C. and M. Meadows (2003) "Framing the future: Indigenous communication in Australia," in N. Couldry and J. Curran (eds) *Contesting Media Power*. Lanham, MD: Rowman & Littlefield, 71–88.

Morris, N. and S. Waisbord (2001) *Media and Globalization: Why the State Matters*. Lanham, MD: Rowman & Littlefield.

Naficy, H. (1993) *The Making of Exile Cultures: Iranian Television in Los Angeles*. Minneapolis, MN: University of Minnesota Press.

NoBorders Network (2010) *Callout for Two Days of Action against Racist Press*, 6 June. NoBorders Network, http://noborders.org.uk/node/19.

Nyers, P. (2003) "Abject cosmopolitanism: The politics of protection in the anti-deportation movement," *Third World Quarterly*, 24(6): 1069–1093.

Nyers, P. (2006) *Rethinking Refugees: Beyond States of Emergency*. New York, NY: Routledge.

O'Kane, M. (2005) *Borderlands and Women: Transversal Political Agency on the Burma–Thailand Border*, Working Paper 126. Victoria, Australia: Centre of Southeast Asian Studies, Monash University.

O'Kane, M. (2006) "Blood, sweat and tears: Women activist-refugees of Burma as political agents," Paper presented at Burma Studies Conference 2006: Communities of Interpretation, National University of Singapore, 13–15 July 2006.

Pries, L. (2009) "Transnationalisation and the challenge of differentiated concepts of space," *Journal of Economic and Social Geography*, 100(5): 587–597.

Sai Silp (2006) "Thai press coverage of Burmese migrant issues criticized," *The Irrawaddy*, 7 July, http://www.irrawaddy.org/aviewer.asp?a=5964andprint=yesandc=e.

Sinclair, J. and S. Cunningham (2000) "Go with the flow: Diasporas and the media," *Television and New Media*, 1(1): 11–31.

Soguk, N. (1999) *States and Strangers: Refugees and Displacements of Statecraft*. Minneapolis, MN: University of Minnesota.

Soguk, N. and G. Whitehall (1999) "Wandering grounds: Transversality, identity, territoriality, and movement," *Millennium—Journal of International Studies*, 28(3): 675–698.

Tyler, I. (2006) "'Welcome to Britain': The cultural politics of asylum," *European Journal of Cultural Studies*, 9(2): 185–202.

Tyler, I. (2008) "'Meanwhile at the borders' … ," *ASA Globalog*, 10 June, http://blog.theasa.org/?p=69.

Walker, R. B. J. (1993) *Inside/Outside: International Relations as Political Theory*. Cambridge: Cambridge University Press.

Walker, R. B. J. (1995) "International relations and the concept of the political," in K. Booth and S. Smith (eds) *International Relations Theory Today*. University Park, PA: Pennsylvania State University Press.

Chapter 9

Anti-Politics and Information Societies in the South

Paula Chakravartty

Overview

Over the last two decades, information and communications technology (ICT) has significantly affected economic prosperity in India. It has also been contributing towards social change. ICT facilitates knowledge-sharing and helps to generate employment, disperse skills and knowledge, and empower the common man. It also expands and enlarges the reach of social programs to all corners of the country at a much-reduced cost.

The National Association of Software and Service Companies (NASSCOM) Foundation (or NF) is harnessing ICT effectively to spread knowledge, increase employability, eradicate poverty, bridge the gender divide, and dispel isolation. NF has two cross-cutting functional areas: to facilitate and strengthen corporate social responsibility (CSR) within the IT–BPO (information technology–business process outsourcing) industry and technology access programs (www.nasscomfoundation.org/index.php/About-Us/Overview.html).

The Conscientious Corporation[1]

As access to and competence in new information and communication technologies has become a defining feature of modern citizenship in much of the global South, it becomes crucial to understand the complex and ambivalent discourse around public–private partnerships and digital inclusion. Drawing from a transcultural approach to the political economy of global communication (Chakravartty and Zhao 2008), in this chapter, I assess both the historical continuity as well as disjunctures in the role of corporate institutional actors in the longer postcolonial history of development communications. Information-infused development projects and the larger technocratic development mandate have a particular resonance in "emergent India." During the last decade, India has become a kind of national "best practice" exemplar of what we can think of as a techno-modernization paradigm associated with the problematic notion of global "digital divide" (Pieterse 2000, 2005; Wade 2002; Lahiri and Pal 2009).

India's importance in this area is partially explained by the symbolic success of its export-oriented information technology industry, often represented as an "engine of growth" for the wider national economy (Chakravartty 2007). The media celebration of the rapid growth associated with the IT industry in info-hub cities such as Bangalore, Chennai, Noida (suburb of New Delhi), and Hyderabad could also be seen as having reinforced the vast disparities between "global" or "shining" India and the growing inequality of everyday life for the vast majority of the nation's citizens (Upadhya and Vasavi 2008). At both national and regional levels, governments closely associated with neoliberal economic reform and IT-led development had lost political power and been replaced by political parties promising reform. At the national level, the United Progressive Alliance (UPA) government headed by the Congress Party that came to power in 2004 implemented a series of reforms meant to extend social equity and welfare under the Common Minimum Program, while retaining an overall commitment to liberalization. Specifically, since 2004, the centrist UPA national government, having extended its economic liberalization mandate in place since 1991, has simultaneously passed historic legislation meant to address both redistribution and accountability in the form of the National Rural Employment Guarantee Act (NREGA) and the Right to Information Act (RTI). Given the dynamic political field, it is important to consider how the prominent IT sector has responded to these new claims for citizenship.

As the website of the prominent IT trade association, NASSCOM, makes apparent, India's new capitalists in the IT sector and their transnational counterparts, most famously Bill Gates of Microsoft, have become passionate advocates of corporate social responsibility (CSR). This often takes the form of "grassroots partnerships" in the arena of development. During the last decade we see a new emphasis on the transformative potential of technologies to empower the "common man" as opposed to the recognized winners of globalization in India: the urban elites. This change reflects a shifting discursive strategy to reposition the IT sector to take on the pressing ethical dilemmas of poverty in the face of neoliberal economic reforms.

Despite the exuberant discourse of economic ascendance, even the conservative World Bank estimates the percentage of India's 1.15 billion citizens living below the poverty line at 42 percent, while other studies put the figure somewhere between 50 to 77 percent of the overall population.[2] In promoting the new future of development communication projects in hundreds if not thousands of local, regional, as well as national programs, corporate actors have been vested in public discourse with both a social conscience and, as a corollary, a language of the rights of citizenship against the powerful and assumed ineffective hand of the state. The Indian IT sector is not unique in this realm; in fact, studies show that corporate philanthropy across the board is on the rise globally in terms of targeting development programs, and that within a very short period of time: "An increasing number of companies are incorporating references to human rights in their policy statements" (Utting 2007, p703). In India, the spectacular nature of economic disparities in the world's largest democracy—an emerging market vying for global recognition with some of the most severe rural poverty rates in the world—has in a sense forced the most visible representative (the private sector) to become an active participant in the discourse of development.

This brings us to the other side of the equation, having to do with the shifting role of the state *vis-à-vis* development and its new strategy of partnering with the private sector, most noticeably perhaps in the arena of ICT-D (information and communications technology development) and e-governance. In this chapter, I want to explore how the emphasis on public–private partnerships (PPPs) incorporates participatory techniques and community/civil society engagement with the objective of promoting "pro-poor pro-market" governance regimes and development projects. These take as a starting point that all pro-poor solutions to the technical problem of poverty alleviation (or closing the digital divide) are amenable to pro-market interventions. Their explicit objectives of promoting state accountability and market transparency have a normative commitment to "depoliticizing" development. In this chapter, I consider the limits of this framework against the shifting political backdrop of greater social and economic inclusion, which has been the hallmark of national and regional politics in the 2000s.

In speaking to the themes of this volume, I would argue that, in many ways, global media studies research has succeeded in decentering our understanding of transnational/local cultural flows and helps us to better make sense of the dizzying explosion of commercial popular culture around the world. But in the global South, as the opening paragraph indicates, a large part of the transformation and rapid proliferation of mediated culture is taking place outside the main purview of the object of study in media studies. Development communication has seen a renaissance in the 21st century comparable to the early decades of the Cold War when winning hearts and minds of the masses in the Third World was intrinsically linked to exposure to modern commercial mass media. Contemporary discussions about the global information society are based on claims about access to "high-tech innovations" vested with the magical potential to transform the lives of individuals and communities in villages, small towns, and urban slums across the "developing world."

As we will see when we look at the new development communication efforts, there are clear differences between the earlier era of geopolitically charged experiments in modernization from above and current iterations emphasizing the organic synergy between corporate benevolence and local community empowerment. Behind the pragmatic and politically neutral tone of the more nuanced discourse of development, however, we see an astonishing acceleration both in the scale of development projects based on information and communication technologies and in its symbolic significance in promising a painless transition to modernity.

Given that "information societies" (emerging or established) are deeply divided as often by neighborhood and regional inequities as by national and transnational borders, I argue that critical media studies must pay greater attention to a parallel and often under-studied terrain in the globalization of the media and information industries: the arena of development communications transformed. In Chakravartty and Zhao (2008, p12) we argue that questions of national identity and modernity are often "relegated to the ghetto of development communication." It is in this context of a shifting global order coupled with a sophisticated revamping of technological determinism where critical political economy and cultural studies approaches would be productive in analyzing both the parameters of policy and governance and the actual practice of technologically driven development.

In making sense of globalization of communications we use the anthropological term "transculturation," a process where the asymmetrical encounter between cultures creates new social and cultural formations. This encounter is constitutive for both non-Western modernities and for the articulation of dominant Western cultural power (Chakravartty and Zhao 2008). New hybrid "partnerships" emphasizing participation in the form of ICT development projects designed and implemented across India challenge our understandings of a global information society imposed from "above" by the heavy hand of the state or "outside" in terms of transnational corporations or foreign interests (Mosse 2005; Ferguson 2006). Correspondingly, the geopolitical stakes of development communication in the 21st century have changed, with the growing influence of emerging powers such as Brazil, China, India, and South Africa in arenas of multilateral governance.

Beyond the shifts in institutional actors, critical media studies scholars are well positioned to assess the new lexicon around technology and governance. Moreover, ethnographic studies of media or, more commonly, studies of technology and development (Escobar 1994; Mitchell 2002; Abu-Lughod 2004; Mosse 2005; Murray-Li 2007; Larkin 2008) offer useful critiques of the limits and logics of participatory approaches and the hype around social entrepreneurship as the panacea to inequality in the global age.

Corporate Social Responsibility (CSR), Development, and the Digital Divide

Similar to changes in the institutional make-up and convergence of global cultural industries, we can identify a series of changes in the field of international development that drew attention to the problems associated with the so-called global "digital divide." As background to these changes, we have to take into account the institutional shifts wherein the object and actors that define state intervention, evident in a new discourse of neoliberal governance, changed fundamentally during the 1980s and 1990s. This meant a shift away from centralized state bodies focusing on domestic performance of the national economy to "partnerships" between private actors, non-governmental organizations (NGOs), and state bodies to coordinate the delivery of social goods and services at the local level. The expanded role of transnational corporations as legitimate participants in shaping social policy, coupled with the taming of civil society actors by incorporating oppositional movements as "stakeholders," is thus the wider political economic context of these changes (Chakravartty and Sarikakis 2006). This reregulation of the state is particularly visible if one contrasts an earlier era of development communication with the current period, where public–private partnerships have become the dominant mode of implementing projects that apply ICTs, including computers and mobile phones, to solve the problems of rural development (Kuriyan and Ray 2009). From its inception in the Okinawa Charter on the Global Information Society in 2000, the formulation of this new techno-developmentalist vision manifested in the new acronym, ICT-D, emphasized the central role of the private sector in ensuring technically expert and efficient delivery of services needed to "bridge the divide" (Chakravartty and Sarikakis 2006, pp51–54).

Before looking at the leading role of the private sector in this new arena of technologically driven development specifically in India, it might be helpful to understand the rise of corporate social responsibility in the context of the challenges of global governance. We also need to account for transnational giants such as Hewlett-Packard (HP), Microsoft, and even Walmart,[3] as well as globally prominent Indian capital, including new IT giants such as Infosys and established business houses such as the TATA group, who became active in this mode of CSR development projects.

Whereas in the 1960s and 1970s there were growing coordinated efforts in the United Nations (UN) to regulate transnational corporations in relation to the developmental goals of nation states in the South, by the 1990s, a new discourse of CSR initiatives emanating largely from Northern social movements (trade unions, environmentalists, human rights activists) became dominant. The fact that in this period most states in the South were in the throes of neoliberal economic reforms meant that CSR priorities were largely defined in negative terms, "with an emphasis on things that companies should not do, such as employing children or violating human rights, rather than on seeking positive development outcomes, such as helping to eradicate poverty" (Jenkins 2005, p528). By the late 1990s, CSR was embraced by the World Bank, as well as prominent national development agencies such as the Canadian International Development Agency (CIDA) and the UK Department for International Development (DFID).[4] The new emphasis on CSR that emerges at this time reflects the crisis of legitimacy within the global governance arena resulting from the dislocating effects of neoliberal economic reforms. Foremost among these effects were the growing rates of inequality both within and across national economies measured in terms of the increasing disparity between society's marginalized and affluent populations.

It is important to recount that at least a decade before the financial meltdown of the US housing market in 2008, inequality and instability in people's everyday lives were jolted by financial crises in Asia and Latin America, reinforcing growing rates of casualization and feminization of labor, growing numbers of displaced peoples and migrants, and the diminished capacity of states to fund health and education and housing programs. In 1999, only a few months after then Senior Vice President and Chief Economist for the World Bank Joseph Stiglitz delivered a speech calling for the end of the "Washington Consensus," Kofi Anan, Secretary General of the UN, addressed 1,000 of the world's corporate elite at the World Economic Forum in Davos, "cautioning against the immanent backlash against globalization, unless it was 'embedded in social values' and reflected 'common objectives' of all segments of the world's population" (Soederberg 2007, p502). In 2000 the UN launched the Global Compact, the world's largest coordinated *but* voluntary CSR initiative that pledges a commitment to good governance, labor, and environmental and human rights principles. According to the UN, the Global Compact—meant to "complement rather than substitute regulatory regimes"—in 2010 "stands as the largest corporate citizenship and sustainability initiative in the world—with over 7700 corporate participants and stakeholders from over 130 countries."[5] The objectives of the Global Compact are meant to "mainstream" the ten principles in business practice and "catalyze" UN objectives in poverty reduction, primarily the Millennium Development Goals (MDGs). Critics point out that, in practice, the Global Compact serves to promote foreign direct investment as the main objective and even an example of CSR (Jenkins 2005, p531). Moreover, a wide range of critical researchers have shown that the Global Compact does little to hold corporations accountable, in practice, providing instead a UN-sanctioned platform for public relations and networking opportunities to promote the growing corporate influence over both national governments and the UN itself.[6]

Nevertheless, for the purposes of this chapter, I would draw our attention to the importance that the Global Compact places on promoting the UN Millennium Development Goals, where for the first time in the UN's history access to information and communication technologies was recognized as a basic human need central to the alleviation of poverty and hunger. The skewed growth of the first stages of neoliberal reform during the 1980s and 1990s—accelerated by the deregulation and liberalization of national telecommunications markets—had very visibly failed to deliver the promise of development. Whereas corporations focused only on their "bottom line" would be unable to meet "social objectives" that served the longer-term interests of both business and the poor, "socially responsible business" could be encouraged or "harnessed" to mobilize pro-poor, pro-market development solutions. Therefore, major institutional players in the development field, from the DFID to the Inter-American Development Bank, the World Bank, and the UN, came to see "CSR as a way of reconciling support for private enterprise and a market-based system with their central aim of reducing global poverty" (Jenkins 2005, p530). Thus, the elevation of information technology as a basic need has to be seen in light of the reframing of development within the UN as well as national development organizations. This vision was built on the existing consensus within these dominant organizations about a "decline in the confidence of the role of the state as an agent for development," as well as the move away from an emphasis on economic growth to a framework defined by the "social dimensions of development" exemplified by the MDGs (Jenkins 2005, p529).

These features of the new development agenda were prominently on display in the UN-sponsored World Summit on Information Societies (WSIS) that took place between 2003 and 2005. While the core issue of financing the "bridge to the digital divide" was left largely unsolved after the UN summit, one lasting impact for countries in the global South was the focused energy on ICT-D as a development priority through the proactive intervention of corporate actors (Chakravartty and Sarikakis 2006; Chakravartty 2007).

This is perhaps most clearly outlined by the World Bank's Infodev program, which was established in 2005 as a multilateral coordinating body for ICT-D-based research and evaluation. As outlined on their website, "*info*Dev's mandate is to help maximize the impact of ICTs in global efforts to achieve the internationally-supported Millennium Development Goals." Its three areas of focus are "Enabling Access for All," "Mainstreaming Access to ICT as Tools for Development and Poverty Reduction," and "Innovation, Entrepreneurship and Growth," by which they mean:

> *info*Dev helps developing countries and the international partners maximize the contribution and impact of the private sector through direct support for ICT-enabled innovation, new business and partnership models and toolkits, and networking among entrepreneurs, private sector investors and the donor community. Our network of over 40 incubators around the developing world provides unique insight into the challenges facing ICT innovation.
>
> World Bank Infodev program (http://www.infodev.org/en/Page.About.html)

What is apparent here is the technocratic mandate that clearly erases the possibility of antagonism between corporate actors and the recipients or, more accurately in this language, participants of poverty alleviation programs and projects. While we can certainly acknowledge the ways in which greater market-based access to mobile phones, for example, has changed the lives of economically marginalized people living in the global South, this does not negate the reality that, in fact, the poor are rarely actual "stakeholders" in this new mode of development. As Jenkins (2005, p540) writes bluntly:

> Almost by definition, the poor are those who do not have a stake. When a TNC invests in a country's capital city, the poor in remote rural regions will not be considered stakeholders; moreover, the decision of many investors to concentrate in the capital both reproduces their poverty and excludes them from being stakeholders.

Transnational corporate actors, including those in the telecom and IT industries, have become key players in both local and transnational efforts in changing the very terms of development. We can therefore track a change in the institutional actors who design, fund, and implement projects—whether promising to reform public education or health through digital technologies, expand e-skills and literacy to create future generations of workers and consumers fueling economic growth, or the very re-imagination of transparent e-governance all together.

As I have tried to argue in this section, the remaking of corporate actors as proactive, conscientious partners in development, especially in the new field of ICT-D, should be assessed critically, and as I will try and establish, should also be located historically. The new language of corporate citizenship so deeply embedded in institutional development discourse today speaks to both the remaking of the state's role in regulating the economy, but also in the reformulation of corporate actors in relation to civil society and political engagement. As Robert Jenkins has pointed out, the "NGO-ization" of the political landscape embraced by development donors tends to purposely conflate social movement criticisms of the violence of the development state with neoliberal economic orthodoxy. In nations like India, with histories of vibrant social movements often at odds with both powerful state actors and elite private interests, civil society organizations are encouraged to mobilize behind "free market" reforms that "removes many decisions not only from the purview of the state, but also the political community, democratically constituted or otherwise" (Jenkins 2005, pp262–263).[7]

Anti-Politics at the Top of the Pyramid

> C. K. Prahalad argues that companies must revolutionize how they do business in developing countries if both sides of that economic equation are to prosper. Drawing on a wealth of case studies, his compelling new book offers an intriguing blueprint for how to fight poverty with profitability.
>
> Bill Gates, Chairman and Chief Software Architect, Microsoft[8]

> The spread of wireless devices, PC Kiosks and personal digital devices at the BOP [bottom of the pyramid] has surprised many a manager and researcher.
>
> Prahalad (2006, p15)

> The BOP consumer is more willing to adopt new technologies because they have nothing to forget.
>
> Prahalad (2006, p16)

In many ways, C. K. Prahalad's 2006 bestselling *Fortune at the Bottom of the Pyramid: Eradicating Poverty through Profits* most effectively captures the discourse of technocratic development in India's ICT-D field. A well-known professor of management at the University of Michigan known in policy circles from the World Bank to the World Economic Forum, Prahalad (2006, p49) argued that "By stimulating commerce and development at the bottom of the economic pyramid, MNCs could radically improve the lives of billions of people and help bring into being a more stable, less dangerous world." Technology is central to this equation in creating both potential entrepreneurs, but also in fueling the aspirations of potential consumers in rural India, which accounts for much of the empirical evidence for his larger argument. One of the seemingly radical gestures that emerges in these interventions, whether one looks at the optimistic media coverage of info-development, state-based policy debates, or industry discussions of CSR in this arena, is the recurring emphasis on reversing an existing failed system of paternalistic state-led development.

As observers have noted, capital in India today "is no longer the privileged bastion of a few mercantile castes" (Damodaran 2008, p312). This presumed widening of the "social base" of a more diversified capitalist class, especially in South India, the epicenter of India's IT industry, has created a resonant market populism that is evident across policy and popular debates.[9] In contrast to the "Brahmanic socialism" of the postcolonial state, the new transnational entrepreneurs, especially in the symbolically important IT industry, are positioned as confident advocates of a techno-populist mandate for middle-class mobility (Chakravartty 2001; Upadhya 2009). During the 1990s, India witnessed the birth of a new transnational capitalist class as familiar icons of Indian modernity whether they were based in Bangalore, such as Wipro's Azim Premji or Infosys founder Nandan Nilekani, or were Non-Resident Indian (NRI) entrepreneurial heroes from Silicon Valley. Following the downturn of the IT industry, this was supplemented by a new discourse of the successful "returning software professional" in cities such as Bangalore and Chennai, which appeared to prove that Indians did not have to remain abroad to be modern (Lahiri and Pal 2009). Pal's recent study points to the fact that the success of the "transnational software professional" began to permeate cinema, politics, and shape the development agenda in the last decade.

Today, at the level of policy, entrepreneurial gurus turned government-appointed cabinet ministers such as Nandan Nilekani are influential voices in both national and global debates on "rethinking development," particularly in relation to the role of information technologies in spurring (transforming) rapid "leaps" to modernization.[10] Nilekani's (2008) proposition that India can only "reap the demographic dividend" of its large young population by extending the opportunities for education, health, roads, electric power, and entrepreneurship to the majority seems innocuous enough. However, his vision of democracy firmly precludes the distributive role of the state, and instead celebrates the autonomous sphere of civil society—the flowering of technically sophisticated civil society organizations working on behalf of the majority of Indian citizens holding state institutions accountable—as the only means for the majority of Indians to meet the aspirational goals of modernity. Growing inequality as a result of the economic reforms in the early 1990s is recast in this discourse as an opportunity for rapid inclusion premised on the non-antagonistic relationship between capital and India's often unruly citizens. This becomes apparent in any cursory overview of the expanding field of ICT-D in India during the last decade where there is broad policy consensus across state and corporate actors on "pro-poor/pro-market" objectives (ensuring state accountability and enabling entrepreneurship) and institutional arrangements (public–private partnerships).

What this has meant, in practice, is the conspicuous absence of public debate about inherently controversial development interventions, reproducing what anthropologist James Ferguson described as the development apparatus operating as an "anti-politics machine." By anti-politics, Ferguson (1994, p270) argued that through the institutional practices of development intervention "political questions of land, resources, jobs, or wages" were reconfigured as technical "problems" responsive purely to "technical development interventions." These include the rapid privatization of land as in the case of Bhoomi in Karnataka, the commercialization of agriculture as in the case of E-Choupal spanning ten states, and investment in social entrepreneurship as in the large national experiment in establishing 100,000 common service centers (CSCs).[11] Ethnographic field-based studies of several prominent e-governance and tele-center projects have demonstrated how varying techniques of de-politicization of the process of development have, if anything, reinforced pre-existing hierarchies in rural Indian society concentrating power in the hands of the already privileged (Benjamin et al. 2006; Sreekumar 2007; De 2008; Gurumurthy 2009; Lahiri and Pal 2009).

I am not arguing that this negative outcome is predetermined because of technological intervention (Kuriyan et al. 2008), nor would I want to deny, as Kuriyan and Ray (2009, p7) claim based on their study of tele-centers in Andhra Pradesh and Kerala, that "public-private tele-centers have created a space for the state to renegotiate its role and image in public service delivery." Clearly, access by marginalized communities to state records in order to ensure the implementation of social programs for social security, food, water, and employment are crucial dimensions of state accountability that can be facilitated through digitization schemes, tele-centers, and greater access to mobile media. However, these necessarily politicized struggles over the implementation of the Right to Information Act (RTI) are distinct from the dominant policy consensus across corporate and state ICT-D actors as they pursue "social entrepreneurship and good governance" (Baviskar 2007; Thomas 2009).

In terms of objectives, there is broad policy consensus that greater access to information reduces both inefficiencies and corruption, exemplified by projects that digitize land records or provide direct information about market prices for agricultural goods. Similarly, there is broad consensus about the compatibility of social development goals of ICT-D projects that are meant to simultaneously address rural inequality through "entrepreneurial" success, as is the case with the massive common service center. Finally, the state both nationally and regionally has aggressively promoted a public–private partnership model in this area, taking as a starting point that private-sector participation is key to sustainability and long-term success. Given the growing influence of corporate actors on economic governance in India, as discussed in the previous section, it is not surprising that critics have raised concerns about the growing convergence in policy vision in the technology-driven development sector.

While corporate actors have become increasingly active in this field, it was initially technologically savvy bureaucrats, primarily from states in South India (Karnataka, Kerala, Andhra Pradesh, and Tamil Nadu), who took the lead in promoting computerization projects meant to make state bureaucracies more accessible to citizens (Lahiri and Pal 2009). Lahiri and Pal (2009, p8) have noted that "visionary IT-Friendly" bureaucrats in charge of projects that would become "best practice" exemplars nationally and internationally "were frequently courted by opportunities outside the government" in the private sector, multilateral organizations, as well as different levels of government. The convergence of these institutional interests in the ICT-D field has meant that despite the fact that it is widely understood that most e-governance projects in the "developing world" fail (Heeks 2002, 2008), these projects continue to be celebrated in media and public policy circles for providing state accountability and fairer market access for the poor.

A case in point is Bhoomi, Karnataka's expansive computerized land-record program (affecting some 5.7 million landowners across 27,000 villages in Karnataka), spearheaded by Rajiv Chalwa, beginning in 2001, who would go on to become the state's e-governance secretary for over a decade (Lahiria and Pal 2009). Recognized internationally as a "success story," its objectives of good governance holding state corruption in check are made clear on its website:

> Bhoomi (meaning land) is the project of on-line delivery and management of land records in Karnataka. It provides transparency in land records management with better citizen services and takes discretion away from civil servants at operating levels.
>
> Bhoomi (http://www.bhoomi.karnataka.gov.in/about.htm)

Lauded by both local and transnational media, Bhoomi's "revolutionary" potential is described in a popular online site, Rediff India Abroad, in the following way:

> Bhoomi means that 7 million farmers in Karnataka can get any extract of land records for a flat Rs 15 in the quickest possible time, sans payment of 'speed money'. It is a network of over a thousand offices (including the latest privately run centres), linked via internet to a central

> depository to which all earlier physical land records have been digitised and transferred.
>
> Rediff India Abroad (http://www.rediff.com/money/2007/apr/10spec1.htm)

Media coverage of projects such as Bhoomi tend to echo the publicity materials provided by the organizations themselves more often than not. Media coverage generally affirms that the projects are administered by "forward looking" bureaucrats (Lahiri and Pal 2009) and embody participatory "citizen-centric" programs meant to empower the poor both politically in terms of state accountability and economically in enhancing entrepreneurial as well as consumption capacity. This takes place despite the fact that it is widely acknowledged by the experts themselves that failure rates are high for e-governance projects and tele-centers alike in India and the "developing world" more broadly (Heeks 2002; Toyoma et al. 2004).

In the case of Bhoomi, an independent report commissioned by the state actually found substantial evidence that small landholders had lost out to large landholding interests after the digitization of land records (De 2008; Lahiri and Pal 2009). Independent research on the Bhoomi's operation in practice by two separate studies found that in certain locales the project tended to reinforce hierarchies of caste and class in landownership (De 2008), while easing the terms of privatization of rural land for large investors (Benjamin et al. 2006). Urban studies scholar Solomon Benjamin and his colleagues have gone on to argue that this process facilitated a "landgrab" by the powerful IT industry based in Bangalore (Benjamin et al. 2006, 2008). Instead of whetting the 24-hour news channels' voracious appetite for stories about unscrupulous land acquisition—a recurring theme in news coverage in contemporary India—the distribution of the report was actively blocked by the bureaucratic body in charge of the project (Lahiri and Pal 2009).

If Bhoomi is one of India's best-known e-governance success stories, e-Choupal (e-village square) is another internationally recognized success story, this time featuring the social entrepreneurship model of ICT-D. Publicity for the project is spelled out on its website featuring visuals that include the familiar motif of smiling farmers surrounding computer screens:

> Today 4 million farmers use e-Choupal to advantage—bargaining as virtual buyers' co-operatives, adopting best practices, matching up to food safety norms. Being linked to futures markets is helping small farmers to better manage risk. e-Choupal has been specially cited in the Government of India's Economic Survey of 2006–7, for its transformational impact on rural lives.
>
> e-Choupal (http://www.itcportal.com/sustainability/lets-put-india-first/echoupal.aspx)

e-Choupal is a private-sector initiative by ITC with diversified interests ranging from cigarettes to hotels to agri-business, ranked as India's tenth most valuable transnational brand with turnover of over US$5 billion (Dangi and Singh 2010). According to C. K. Prahalad (2006, pp325–329), where e-Choupal featured prominently as a best practice model of social entrepreneurism, ITC's success in creating a "pro-poor/pro-market" formula that works is based on the company's commitment to the "social agenda" and its dedication to "nation building." The scale of the project across ten states and serving some 40,000 villages, the largest operational ITC-D project, is in itself impressive. While there is evidence that the project has increased rural agricultural incomes by introducing "efficiency improvements in the procurement chain," the distributional impact of these benefits has been uneven and, like Bhoomi, suggests that they might, in fact, reinforce inequalities for the majority of small landholders (Dangi and Singh 2010). Moreover, critics have pointed out that the initiative seeks to become the "Walmart of rural India" (Gurumurthy 2009), with essentially "monopolitistic control over the entire local agriculture ecology by a transnational corporation through the use of a captive ICT infrastructure with minimal regulation and competition" (Dangi and Singh 2010, p182). Based on a business model providing connectivity and services to a closed network of farmers through an entrepreneur whose role, interestingly, is projected by ITC as a "public office" (Prahalad 2006, p179), e-Choupal exemplifies the win–win logic of this new development discourse (Gurumurthy 2009).

This brief and critical overview of projects that are both locally and internationally lauded for their "empowerment" of poor farmers in India is meant to draw attention to the limits of the "anti-politics" turn in information-based development in India and beyond. Empirical research on ICT-D in India is extensive, and the limits of much of the policy-driven analysis on development outcomes have already been spelled out by a range of critics, from scholars (Sreekumar 2007; Schwittay 2008; Lahiri and Pal 2009; Thomas 2009) to NGO-based activists (Benjamin et al. 2006; Gurumurthy 2009) and even from collaborations between academics and researchers based in corporate think-tanks (Kuriyan et al. 2008; Burrell and Toyoma 2009).

My purpose in bringing up these particular exemplary ICT-D projects, lauded by a variety of national and international foundations, is to simply show the obvious contradictions between the "pro-poor/pro-market" framework that forms the normative basis for the vast majority of initiatives in India. These modernization schemes promised to democratize the benefits of India's skewed information society. And within a decade, "pockets" of rural India became a "Living Lab" for public–private partnership based delivery of information technology, promising nothing short of social transformation (Schwittay 2008).

To give some sense of the contemporary scale of this sector, one independent report estimated in 2007 that there were between 12,000 and 13,000 tele-centers or cyber-kiosks operating in the country, with approximately 50 percent as state operated or public–private partnerships, and the rest as private for-profit initiatives in areas such as e-agriculture (APC 2007, p158). The Government of India is currently spending 230 billion Indian rupees (or between 2 to 3 percent of its overall budget) on its National E-Governance Plan. The centerpiece of the scaled-up project of national e-governance, announced in 2006, was a plan to establish 100,000 common service centers (CSCs) across India's 600,000 villages. The objective was to establish a "multi-stakeholder" program that allows "the common man" greater access to government services and the opportunity for "rural entrepreneurship." Although the Department of Information Technology estimates that some 80,000 of these CSCs are operational as of 2010, emphasizing the "success stories" of hundreds of "village-level entrepreneurs" who provide the "vital link" to remote rural communities, literally embodying the bridge to the so-called "digital divide," ongoing critical exploration of these projects on the ground is likely to tell different stories.[12]

Conclusions

During the last decade, India has become a thriving site for local and regional case studies of experiments in techno-development evaluated and celebrated by foundations and multilateral institutions. Design, implementation, and evaluation of these dizzying numbers of localized development experiments has created an army of experts in e-governance and grassroots digital inclusion efforts who are tirelessly collecting survey and ethnographic data, presenting PowerPoints, and meeting in what has become an institutionalized succession of workshops, conferences, and symposiums where the technologically enhanced rights of the newly empowered citizen are mapped, charted, and recalibrated for the next cycle of projects in India and well beyond. The Indian government, multilateral funding agencies, corporate actors, and civil society organizations see greater access to new communications technologies as a necessary "bridge" capable of overcoming disparities, ensuring accountability, and ultimately reversing inequality. In this chapter I have traced the historical continuities and discontinuities in the role of corporate institutional actors within a longer postcolonial history of development in India, following a transcultural approach to the political economy of ICT-D (Chakravartty and Zhao 2008).

First, I have argued that the current public–private institutional convergence is a coherent vision of national development in the longer history of 20 years of social contest over India's emergent unequal information society. The role of the private sector has to be assessed against the background of the growing calls for social and economic inclusion, which marks the last decade of liberalization in India. The transnational IT industry elite in India, who became iconic representatives of a new Indian capitalism, as discussed above, had a "disproportionate impact on public policy" both through national business associations and transnational business networks (Upadhya 2004). As industry insiders themselves point out, NASSCOM (the national IT trade association) and transnational IT industry in India have been extremely successful in "wresting most concessions they want from government," including stronger intellectual property rights (IPR) regimes and favorable tax and labor and infrastructure policies, while presenting themselves as vocal critics of government-led development and institutionally autonomous from state power (Atheye and Chaturvedi 2007, p25).

Just as a reflection of neoliberal shifts in governance, national states are learning from and behaving more like transnational corporations, while transnational corporations and national corporate stars alike invest significant sums on public relations efforts to prove that they are, in fact, good citizens. Corporate social responsibility (CSR) in this context can be seen as an important marketing strategy employed by the private sector engaging especially in extractive, exploitative, and highly unequal economic sectors in the global South, including the oil industry, the garment industry, and the computer and electronics industry (Watts 2005). As Knight and Smith (2008, p9) have argued: "By casting this role in the language of responsibility rather than obligation, CSR advocates soften both the attribution of causal blame for social and environmental problems and the strength of any normative requirement on TNCs [transnational corporations] to assume the costs and risks of remedial action." A glaring example has to do with the ways that CSR schemes, like those discussed above, neglect one conspicuous area which seems fundamental to good corporate citizenship: increasing corporate tax burden to contribute state resources, allowing for more effective delivery of public services such as education and health.

While private-sector philanthropy in absolute terms is relatively low in India, the influence of the private sector in increasingly shaping development priorities in terms of urban planning, health, education, and employment generation has increased substantially in the last decade. Corporate citizenship is certainly the dominant trend in the high-profile IT industry, whether we consider companies such as Dell Computers and the HP Foundation, the Microsoft Foundation, the NASSCOM Foundation, the Azim Premji Foundation, and hundreds of others where firms and their foundations play a driving role in public–private partnerships in e-governance, or the myriad kiosk and now mobile media-based ICT development projects. In the last section of this chapter I discussed some of the general trends and provided brief overviews of two of the most successful projects. The contradictions of the pro-poor/pro-market logic that informs projects such as Bhoomi in Karnataka and e-Choupal across ten states are instructive of the limits of projects whose guiding vision is limited to *reframing the poor* as more active participants in state accountability and empowered entrepreneurs. In contrast to an earlier vision that "characterized the poor largely as passive consumers, the next phase of 'ICT4D 2.0' promises to reframe the poor as active producers and innovators" (Heeks 2008, p32).

For countries such as India with histories of state-led development and, in its case, the legacy of the "bureaucracy raj," nation branding needs to establish a clear rupture from the previous era of state intervention. The promise of a "new" India is premised around the recuperation of the individual citizen/consumer through the strengthening of non-state actors or civil society, meaning NGOs and corporations. As anthropologist Charles Hale (2002, p496) has argued in the context of Latin America, neoliberalism is based on a critique of state intervention (the welfare state, developmentalist state, etc.) and "is predicated on the need to recreate or recapture the individualist essence, in danger of being lost." If the initial stage of neoliberal reforms in India promised simply to unleash the entrepreneurial instincts of all Indians, in the current stage of "inclusive" liberalization public–private partnerships offer particular attention to the citizen-subjects thus far left out of the benefits of global India. A focus on neoliberal reform and "inclusive growth" usually then means the neoliberal state "unload[ing] onto its neoliberal citizen-subjects the responsibility to resolve the problems—whether daily or epochal—in which they are immersed" (Hale 2002, p496).

Notes

1 I am using a term that is developed by Christina Gartsen and Tor Hernes (2009, p197) in distinguishing a new turn in how "corporate actors position themselves in society and rework organizational boundaries." I will discuss this further in the chapter.
2 See http://www.im4change.org/articles.php?articleId=40&pgno=2.
3 Lahiri and Pal (2009) have argued that, increasingly, "a sizeable number of companies that do not make their money from technology are nonetheless including what one may consider 'ICTD' projects in their CSR funding profiles." These include Citigroup, ExxonMobil, as well as Walmart.
4 For more detailed discussion of the CSR agenda as embraced by development agencies, see Jenkins (2005).
5 See http://www.unglobalcompact.org/AboutTheGC/index.html.
6 For an updated list of published research by both academics and civil society activists on the Global Compact, see http://globalcompactcritics.blogspot.com/#uds-search-results.
7 The history of civil society engagement in debates about the global information society, and especially the politics of information societies in the global South, is discussed in greater detail in Chakravartty (2007).
8 The quote from Bill Gates is an endorsement that appears on the book jacket for C. K. Prahalad's popular book, *Fortune at the Bottom of the Pyramid* (2006). Tellingly, the other endorsement comes from Madeline Albright, former US secretary of state, demonstrating the convergence in corporate and state "commonsense" when it comes to this new take on development:

> *The Bottom of the Pyramid* belongs at the top of the reading list for business people, academics, and experts pursuing the elusive goal of sustainable growth in the developing world. C. K. Prahalad writes with uncommon insight about consumer needs in poor societies and opportunities for the private sector to serve important public purposes while enhancing its own bottom line. If you are looking for fresh thinking about emerging markets, your search is ended. This is the book for you.

9 It is important to point out that the issue of greater diversity across India's liberalized private sector is a source of controversy, with empirical studies showing the reverse to be true, for example, in Bangalore's IT industry (Upadhya and Vasavi 2008). And Harish Damodaran (2008, p315), who makes the case for greater diversity in caste backgrounds of "India's new capitalists," finds that "The group that has ... been bypassed in this churning of capital and widening of its social base are the Dalits."

10 In 2009, Nilekani became a cabinet member in the ruling UPA government as chair of the Unique Identification Authority of India, the controversial national program to issue identity cards based on biometric data. Advocates claim the program will allow the poor better access to services, while critics worry about civil rights and greater surveillance of already marginalized communities.

11 Benjamin et al. (2008), De (2008), and Thomas (2009) all discuss the case of Bhoomi. e-Choupal is discussed in Gurumurthy (2009) and Dangi and Singh (2010). My analysis of the CSC project is based on my own ongoing fieldwork in Delhi, Andhra Pradesh, and West Bengal since 2008.

12 Current figures given by the Department of Information Technology can be found at http://www.mit.gov.in/content/common-services-centers. On the same page is also a link to a report on *CSC Success Stories*, highlighting individual experiences of village-level entrepreneurs.

References

Abu-Lughod, L. (2004) *Dramas of Nationhood: The Politics of Television in Egypt*. Chicago, IL: University of Chicago Press.

APC (Association for Progressive Communication) (2007) *Global Information Society Watch Annual Report*, http://www.giswatch.org/gisw2007/.

Athreye, S. and S. Chaturvedi (2007) "Industry associations and technology-based growth in India," *European Journal of Development Research*, 19(1), http://www.ilo.int/public/english/bureau/inst/papers/confrnce/research/sumapaper.pdf, accessed 15 March 2009.

Baviskar, A. (2007) *Is Knowledge Power? The Right to Information Campaign in India*. London: Institute of Development Studies.

Benjamin, S., R. Bhuvaneshwari and P. Rajan (2006) *"E-Titling" of an "Anti-Politics Machine" to globalize Bangalore*, International Conference on Enhancing Land Registration Cadastre for Economic Growth in India. New Delhi: Center for Science, Development and Media Studies.

Benjamin, S., R. Bhuvaneshwar, and R. Rajan (2008) "Fractured terrain, spaces left over or contested: A closer look at the IT-dominated territories in South and East Bangalore," in D. Mahadevia (ed) *Inside Transforming Urban Asia: Processes, Policies and Public Actions*. New Delhi: Concept Publishing, 239–285.

Burell, J. and K. Toyoma (2009) "What constitutes good ICTD research?," *Information Technology and International Development*, 5(3): 82–94.

Chakravartty, P. (2001) "Flexible citizens and the Internet: The global politics of local high-tech development in India," *Emergences: Journal for the Study of Media and Composite Cultures*, 11(1): 69–88.

Chakravartty, P. (2007) "Governance without politics: Civil society, development and the postcolonial state," *International Journal of Communication*, 1: 1–22.

Chakravartty, P. and K. Sarikakis (2006) *Globalization, Communication and Media Policy: A Critical Perspective*. Edinburgh: Edinburgh University Press.

Chakravartty, P. and Y. Zhao (eds) (2008) *Political Economy of Global Communication: Towards a Transcultural Perspective*. Boulder, CO: Rowman & Littlefield.

Damodaran, H. (2008) *India's New Capitalists: Caste, Business and Industry in a Modern Nation*. New York, NY: Palgrave MacMillan.

Dangi, N. and H. Singh (2010) "e-Chopal: Hope or Hype?," *American Journal of Economics and Business Administration*, 2(2): 179–184.

De, R. (2008) "Control, de-polititicization and the eState," *Lecture Notes in Computer Science* (LNCS), 5184: 61–74.

Escobar, A. (1994) *Encountering Development: The Making and Unmaking of the Third World*. Princeton, NJ: Princeton University Press.

Ferguson, J. (1994) *The Anti-Politics Machine: Development, Depoliticization and Bureaucratic Power in Lesotho*. Minneapolis, MN: University of Minnesota Press.

Ferguson, J. (2006) *Global Shadows: Africa in the Neoliberal World Order*. Durham, NC: Duke University Press.

Fuller, C. J. and H. Narasimhan (2007) "Information technology professionals and the new rich middle class in Chennai (Madras)," *Modern Asian Studies*, 41(1): 121–150.

Gartsen, C. and T. Hernes (2009) "Beyond CSR: Dilemmas and paradoxes of ethical conduct in transnational organizations," in K. E. Browne and B. L. Milgram (eds) *Economics and Morality: Anthropological Approaches*, Plymouth, UK: AltaMira Press, 189–210.

Gupta, A. (1998) *Postcolonial Developments: Agriculture in the Making of Modern India*. Durham, NC: Duke University Press.

Gurumurthy, A. (2009) *Social Enterprises to Mobiles: The Curious Case of Propped-Up ICTD Theory*. Harvard: Berkman Center for Internet and Society, http://publius.cc/social_enterprise_mobiles_%E2%80%93_curious_case_propped_ictd_theory/091709, accessed 18 January 2010.

Hale, C. (2002) "Does multiculturalism menace: Governance, cultural rights and the politics of identity in Guatemala," *Journal of Latin American Studies*, 34: 485–524.

Heeks, R. (2002) "Information systems and developing countries: Failures, successes and local improvisations," *The Information Society*, 18(2): 101–112.

Heeks, R. (2008) "ICT4D2.0: The Next Phase of Applying ICT for International Development," *Computer*, 41(6): 26–33.

Jenkins, R. (2004) "Mistaking governance for politics: Foreign aid, democracy and the construction of civil society," in S. Kaviraj and S. Khilnani (eds) *Civil Society: History and Possibilities*. Cambridge: Cambridge University Press, 250–268.

Jenkins, R. (2005) "Globalization, corporate social responsibility and poverty," *International Affairs*, 81(3): 525–540.

Knight, G. and J. Smith (2008) "The global compact and its critics: Activism, power relations and corporate social responsibility," in J. Leatherman (ed) *Discipline and Punishment in Global Politics: Illusions of Control*. New York, NY: Palgrave MacMillan, 191–214.

Kuriyan, R. and J. Ray (2009) "E for express: 'Seeing' the Indian state through ICTD," *IEEE/ACM Proceedings, Information and Communication Technology and Development*, http://docs.google.com/viewer?a=v&q=cache:n3TRzvxP59IJ:tier.cs.berkeley.edu/tierwiki/media/a/a9/ReneeKuriyan_IshaRay_ICTD2009.pdf+E+for+Express:+%E2%80%9CSeeing%E2%80%9D+the+Indian+State+through+ICTD%E2%80%9D&hl=en&gl=us&pid=bl&srcid=ADGEESgeJBc_7XlZFQZ4gelTc7TRvqjwb4G_wZhHWlvItoj9V99_b91VEBnrf26_yPNQOak5xuqjzNmAWF

OHElkWJCL1diLGNQ6X2582rjCpXlFdOuwpiw5kNSOeMvjuvSIH5xhb1FXD&sig=AHIEtbSdzuutfSd6lmjCDu2yoYymMY1uMg, accessed 15 January 2010.

Kuriyan, R, I. Ray, and K. Toyoma (2008) "Information and communication technologies for development: The bottom of the pyramid model in practice," *Information Society*, 24: 1–12.

Lahiri, A. and J. Pal (2009) "ICTD in corporate social responsibility: Changing priorities in international development funding?" *Proceedings IEEE/ACM Proceedings, Information and Communication Technology and Development*, http://docs.google.com/viewer?a=v&q=cache:zPnaKsOfTIUJ:www.globdev.org/files/proceedings2009/26_FINAL_Lahiri_ICTD_in_Corporate_Social_Responsibility_2009.pdf+ICTD+in+Corporate+Social+Responsibility:+Changing&hl=en&gl=us&pid=bl&srcid=ADGEESgV685Hs_M_5wa1bTszVkw_i5Tv57TGDYrhAV7KyvO3SNDHRJxqUE6yF3qbYPTCezjo5sVHt-wJem06m5e5aIml-tX1XMUDYNyI-lPrA60aN8ATgSNUwoXFtBB-AcZJPbAz1Ct1&sig=AHIEtbQ5Y0Lv9NUXYlLQhRF4xAH_pVYRQw, accessed 15 January 2010.

Larkin, B. (2008) *Signal and Noise: Media, Infrastructure, and Urban Culture in Nigeria*. Durham, NC: Duke University Press.

Microsoft Corporation (2008) "Microsoft strives to bring relevant, affordable and accessible technologies to the world's rural underserved populations," 22 July, http://www.gkpeventsonthefuture.org/, accessed 14 November 2009.

Mitchell, T. (2002) *Rule of Experts: Egypt, Techno-Politics, Modernity*. Chicago, IL: University of Chicago Press.

Mosse, D. (2005) *Cultivating Development: An Ethnography of Aid Policy in Practice*. New York, NY: Pluto Press.

Murray-Li, T. (2007) *The Will to Improve: Governmentality, Development, and the Practice of Politics*. Durham, NC: Duke University Press.

Nilekani, N. (2008) *Imagining India: Ideas for the New Century*. Delhi: Penguin.

Pieterse, J. N. (2000) 'Trends in development theory,' in R. Palan (ed) *Global Political Economy: Contemporary Theories*. London and New York: Routledge, 197–214.

Pieterse, J. N. (2005) "Digital capitalism and development: The unbearable lightness of ICT4D," in G. Lovink and S. Zehle (eds) *Incommunicado Reader*. Amsterdam: Institute of Network Cultures, 11–29.

Prahalad, C. K. (2006) *Fortune at the Bottom of the Pyramid: Eradicating Poverty through Profits*. New York, NY: Dorling Kindersley.

Rajak, D. (2009) "'I am the conscience of the company': Responsibility and gift in a transnational mining corporation," in K. E. Browne and B. L. Milgram (eds) *Economics and Morality: Anthropological Approaches*. Plymouth, UK: AltaMira Press, 211–231.

Schwittay, A. (2008) "'A living lab': Corporate delivery of ICTs in rural India," *Science, Technology & Society*, 13(2): 175–209.

Sinha, A. (2005) "Understanding the rise and transformation of business collective action in India," *Business and Politics*, 7(2): 1–35.

Soederberg, S. (2007) "Taming corporations or buttressing market-led development? A critical assessment of the global compact," *Globalizations*, 4(4): 503–516.

Sreekumar, T. T. (2007) "Cyber-kiosks and the dilemmas of social inclusion in rural India," *Media, Culture and Society*, 29: 869–887.

Thomas, P. (2009) "Bhoomi, Gyan Ganga, e-governance and the right to information: ICTs and development in India," *Telematics and Informatics*, 26(1): 20–31.

Toyoma, K. et al. (2004) "Rural kiosks in India," Research Microsoft, http://research.microsoft.com/apps/pubs/default.aspx?id=70150, accessed 9 June 2008.

Tsing, A. L. (2004) *Friction: Ethnography of Global Connection*. Princeton, NJ: Princeton University Press.

Upadhya, C. (2004) "A new transnational capitalist class? Capital flows, business networks and entrepreneurs in the Indian software industry," *Economic and Political Weekly*, 27 November: 5141–5151.

Upadhya, C. (2009) "Imagining India: Software and the ideology of liberation," *South African Review of Sociology*, 40(1): 76–93.

Upadhya, C. and A. R. Vasavi (2008) "Outposts of the global information economy," in C. Upadhya and A. R. Vasavi (eds) *In an Outpost of the Global Economy: Work and Workers in India's information Technology Industry*. New Delhi: Routledge, 9–49.

Utting, P. (2007) "CSR and equality," *Third World Quarterly*, 28(4): 697–712.

Wade, R. H. (2002) "Bridging the digital divide: New route to development or new form of dependency?" *Global Governance*, 8(4): 53–58.

Watts, M. (2005) "Righteous oil: Human rights, the oil complex and corporate social responsibility," *Annual Review of Environment and Resources*, 30: 373–407.

Chapter 10

New Mediations in the Digital Age

An Analysis of Global Communication through Professional Journalists

José María García de Madariaga

Social communication has been transformed by new paradigms involving digitalization: interactivity, customization, the use of multimedia, hypertext links, unlimited amounts of information that may be continually updated, and finally, the gradual transformation of traditional processes of informative mediation, six of the main keys of global communication (Orihuela 2002). However, changes that promote new information and communication technologies (ICTs) in the field of journalism are much more significant than in news broadcasting, increasingly more and more abundant, unmediated and in real time, or "journalism on demand." These phenomena could be, as Ignacio Ramonet (2001, pp49–50) has warned, worrying signs of the "tyranny of communication" prevalent in today's globalized world, in which he says "seeing is understanding": "The system blames the receiver [of information] who can now denounce [misinformation as] lies, because he has become informed on his own."

Such concerns lead us to wonder who benefits from and who is harmed by the digitization of journalism, a process that, while closely related to globalization, is not restricted to the technological aspects of this phenomenon. This question also requires us to understand the magnifying impact digitization has upon the structural imbalances in the production, circulation, and accessibility of information, imbalances that in our globalized era led to the creation of a "digital divide." But it also suggests that digital technologies offer citizens exciting opportunities to engage as potentially significant players in the journalistic arena, possibly redressing the growing imbalance that results from the hegemonic influence of much larger and many fewer media companies.

If increasingly concentrated capital is the biggest beneficiary of this transformational process, its growth has had the effect of a devastating hurricane, leaving behind many victims in the form of media companies, which are now defunct or are suffocated by stiff competition from the major companies, and thousands of journalists left unemployed in an increasingly precarious labor market. The most notable victim of this corporate tsunami, apart from professional journalism, is information itself, which is increasingly of poorer quality and marked by a more limited and homogeneous worldview.

Focus and Objectives

This chapter seeks to identify those factors that affect the quality of the work of journalists in an increasingly interactive digital environment, and to assess to what extent these factors legitimize the role of media professionals in ensuring the accessibility, reliability, and credibility of information.

Our analysis is guided by the hierarchy of influences model proposed by Pamela Shoemaker and Stephen Reese in their book *Mediating the Message* (1996), which serves as an adequate conceptual framework for analyzing the quality of the work of journalists and examining how the media function. This model provides a theoretical framework for an analysis at five levels: individual, operational (the newsroom), organizational, environmental, and ideological. Such a framework for the whole does not imply an acceptance of any deterministic set of positions. Rather, it makes sense to conceptualize a given variable as differentiated into various dimensions, but also sharing between these dimensions the conditions that mark the reality with which they are engaged dialectically.

This study involves three specific goals:

1 Assess the direct and indirect influences of new ICTs in the professional environment of journalists. Despite the fact that many journalists have trouble seeing beyond this operational aspect of technology, we will examine the relationship of journalists to technology, in general, and to the Internet, specifically.
2 Identify what are the most critical traits for journalists to best work in an interactive communication environment. We will seek to identify the degree of subject matter and conceptual expertise required for professionals with these skill profiles to meet the needs of an interactive society in which the new voices on the Internet are profoundly knowledgeable about their subjects.
3 Evaluate the potential of the Internet as a platform for a more collaborative, profound, and professional kind of journalism. We are interested at this point in determining whether the incorporation of digital interactivity in society also leads to adopting interactive methods that will help newsrooms to better adapt to the changing needs of an audience very different from that of previous years, and in a continual state of transformation. In short, we hope to determine which factors make for the best use of the Internet, not only as a professional tool for use by journalists.

This work is based on four sets of assumptions. First, new communication paradigms require a revision of the traditional canons of journalism and of the journalist as a personal and professional agent embedded in a complex environment under the influence of a growing concentration of corporate power converging technologies. On the one hand, this context reinforces the journalist's operational and intellectual capabilities; but, on the other, it restricts him or her to the role of mere disseminator of information that has been selected and even handled outside of his or her control.

Second, professional journalists are not taking advantage of the Internet's potential for their work for several reasons: media and ICT users generally have limited training, and work under pressure due to the business or labor practices and policies of their workplaces. As a result, communication professionals are not aware of or do not value many of the advantages that using the Internet at work could potentially mean for them, particularly as regards their independence professionally.

Third, social movements, increasingly more visible in a society of digitalized information, find more effective ways to be heard, which also involves a remarkable transformation as a result of the incorporation of new ICTs. This can considerably affect the messages produced by journalists within mainstream media. Similarly, traditional media and media professionals may incorporate interactive and collaborative methods to enrich their reporting, just as many initiatives for counter-information will bolster their role and actions by using mechanisms for more open social participation.

Fourth, the influence and responsibility of media entities will continue to play a vital role in maintaining the links and connections necessary for a cohesive social order in spite of the impact upon media operations and upon the functional role of journalists resulting from the appearance of new ICTs and their use by journalists, social movements, and the public, in general.

Journalism in Crisis

If the information worker of the digital age—in which information becomes the principal raw material—is a present-day successor to the steel worker of the industrial age, his or her circumstances are very different than those of the period of industrialism. Today's journalist is the clearest example of intangible work: a cog in the gears of the pervasive technocratic rationalism of our digital society, which highlights the cultural contradictions of capitalism, as Daniel Bell (1989, p89) has suggested.

An endless series of mergers and takeovers has given rise to powerful and modern entertainment and news companies. In turn, this corporate landscape has fueled a growing technological convergence that characterizes huge media conglomerates, which today dominate both the market and culture. Multimedia giants such as the Murdoch Group, AOL–Time Warner, Microsoft, or Google now lead the business of communication, culture, and entertainment. These empires not only control the production of content, but are also present throughout the media chain, especially in content distribution, the real business battleground in this "Age of access," as Jeremy Rifkin (2000) has described it.

Two noteworthy factors affecting news professionals are evident in sociological studies of the newsroom (Johnstone et al. 1976; Gans 1979; Fishman 1983; Tuchman 1983; Shoemaker and Reese 1996) and of the profiles of journalists (Weaver 1998; Canel and Sádaba 1999; García de Cortázar and García de León 2000; Ortega and Humanes 2000). These are, first, the evolution of corporate structures, increasingly globalized and concentrated, and, second, the development of new ICTs—both within and outside of newsrooms, which means a greater capacity to handle information through digitalization (Shoemaker and Reese 1996; Weaver 1998; Boczkowski 2002; Martínez-Nicolás et al. 2006; García de Madariaga 2008). Both factors have been key in identifying and analyzing the characteristics of this particularly diffuse and deregulated activity. Deregulation of markets and growing corporate concentration have weakened the professional journalist's traditional sources of support: the objectivity and the role of the media as monitors of power are less valuable in a society which is increasingly subject to the dictates of commercial rationality (Bruns 2005, p308) and which turns most journalists into mere disseminators of content (Canel and Sánchez-Aranda 1999; Ramonet 2001).

Furthermore, compared to the journalist of the assembly line, citizens today, with access to more and more sources of information, are increasingly interconnected and in many cases actively engaged as "citizen" reporters, capable even of challenging the work of professionals (Sandoval Martín 2005; García de Madariaga 2006). The convergence of both the technology and the business of news, therefore, poses two serious obstacles—almost simultaneously—for journalists to play their assigned professional role in a democratic society. This explains, among other reasons, the increasing job insecurity that has characterized the work of journalists for a number of years (Randall 1999; Ramonet 2001).

Given this context of crisis for both journalists and newsrooms (Wolf 1996), and faced with the need to change their skills and professional practices (Randall 1999; García de Madariaga 2004) to adapt to a work environment where they are no longer seen as having an exclusive hold on their profession, the emergence of new and diverse forms of news dissemination, particularly via the Internet, instead of modifying the professional culture of journalism would appear to have preserved the same codes and professional attitudes that prevailed during the 20th century (Boczkowski 2002, 2004; Martínez-Nicolás et al. 2006). It remains to be seen if the consolidation of online newsrooms will require journalists to adapt professionally or accept changed employment conditions as newsrooms now exist in a digital world.

As a result of this "downgrading" of journalism, a profession which some authors already diagnosed as ill (Pablos Coello 2001) or as endangered (Ramonet 2001), much of the journalist's role in selecting and interpreting news has been displaced to decisions made by boards of directors of media and advertising conglomerates, by lobbyists, and by interest groups. To a lesser extent, and yet remarkably in some cases, this shift is also due to counter-information campaigns, which in a digitalized media environment have gained relative strength and exercised a degree of impact upon public opinion.

Thus, a reporter is limited mainly to conveying information that has been previously slanted and pre-selected by established channels, and will contribute little more than packaging it suitably for consumption by a specific audience. Very few radio and television celebrity journalists and news-anchors, as well as well-known editorial writers, continue to project the mythical image of the journalist as a powerful professional in his own right. This idealized image was not all that realistic, as is known from cases noted by Lippmann (2009), such as the excessive reliance of reporters on government officials and spokesmen in the 1920s. José Luis Dader (1992, p82) has remarked on the "underlying contradiction between the journalist's role as an investigator working on behalf of the public, which presumably is a role assigned to him by a democratic society, and the task of merely reporting on what others say has happened, a role which in practice is played by most of these professionals" (translation from original quotation). This contradiction is only further exacerbated by the new ICTs, as we have already noted. However, although we are now more than ever before able to act quickly, efficiently, and independently, the notable trend of journalism in this age of convergence is increasingly to locate and uncritically convey as news reports the prepared statements of media-savvy spokespersons.

As a result, although journalism as a profession has had the task of reporting on the course of society, this has not resulted in recognition or professional status similar to that of other trades such as law and medicine, even though journalists still enjoy a fairly high social standing. Professional journalism seems to evoke a mythical image more akin to that of adventurers or detectives, who have little in common with the intellectual poverty experienced every day by most journalists. Further still from this glorified image is the dwindling recognition that journalists receive from the marketplace, chiefly responsible for socio-economic classifications in our globalized world. Both of these aspects—the quality of professional work and job security—shape and affect the endangered nature of this profession. The work product of this activity is often not up to the expected standards of quality, and shows worrying symptoms of a profession that is increasingly powerless. Although journalists rarely engage in self-criticism, and much less publicly or openly, they admit being disappointed at the excesses and shortcomings of the profession as it is currently practiced.

An excessive use of dramatization and sensationalism, the overloading of information as a new kind of censorship—without abandoning traditional techniques of not reporting—and the dizzying acceleration of the pace of information are some of the symptoms of a diagnosis referred to as the malady of "infotainment."

In writing about this crisis, many observers have emphasized the role of the media companies rather than the professionals themselves. Experience and empirical studies show that in addition to a reduced elite group of star-journalists and editors, another larger group of professionals is characterized by its diminishing autonomy, and is increasingly subject, on the one hand, to corporate restrictions and, on the other, to the whims of commercialized audiences, which are strictly controlled by rigid corporate hierarchies and are internalized through self-censorship. However, this insecurity does not prevent even rank beginner journalists from considering themselves as belonging to a cultural elite whose highest professional aspiration is to influence society.

Moreover, professional inbreeding does not alone account for the great decline in the quality of information or in the job insecurity felt by journalists. Many different arguments may help to explain this decline. An excessive labor supply, which explains the crowded enrollment of journalism schools, or the persistent uncertainty of the practice of the news profession, tediously manifest in the unending and futile theoretical discussions about what is the definition of a journalist[1] or whether journalism is, in fact, a profession, could all serve as explanations of why the vast majority of news professionals have such unstable employment, and in many cases must resort to moonlighting. However, such explanations, rather than addressing the causes of the problem, represent underlying signs that something is amiss in the profession.

Technological Change

A search for the deeper reasons behind this crisis may be found by assuming that something has changed in the journalistic landscape that increasingly affects all areas of the profession. Beyond the specific and obvious symptoms is a particularly individualistic approach, which results from an all too common misunderstanding of what professionalism entails, and which might account for the extreme and poisonous individualism among journalists. Once again, we must look to changes in technology, which in recent years have been numerous, rapid, and very intense, but more importantly have transcended their role as instruments of the profession. According to Víctor Manuel Marí Sáez (1999, p14):

> This phenomenon is unprecedented in human history, since to date no technology had influenced social life so broadly (economics, politics, culture) nor done so on a global scale. The novelty lies in the all-encompassing nature of this technological revolution: all dimensions of life and global society are affected.

At the turn of this century, millennium, and era, technology is particularly important because of the role it has as a factor in the convergence of change. Until now, new media and new tools brought about more or less isolated changes in the practices and methods of journalism, and even caused structural changes in the development of the profession. This was the case in the late 19th century when typesetting and linotype did away with an older generation of woodblock printers who set type by hand. Later still, the use of photo-type again quadrupled the speed with which printing plates were produced, but also led to numerous social and labor problems due to the significant effect this had on the workforce needed.[2]

These episodes, however, show that as digitization has absorbed work processes and made our globalized economy more compact, the effects of technology are not isolated, but instead are convergent. Similarly, a digital environment allows for shared tools and practices. Just as the Internet is able to compatibly bring together different operating systems, document formats, support, and multimedia applications, the convergence of technology also finds an ideal environment in this highly interconnected world of business and multimedia entities.

Among the many voices sounding the alarm of the predicament of journalism from within the ranks of the profession is that of Ignacio Ramonet, who has diagnosed that the profession is, indeed, ill, and warned that journalists are "in danger of extinction" (Ramonet 2001, p47). The editor of *Le Monde Diplomatique* identifies this as originating in both the technological and economic phenomena that could be understood as a "second capitalist revolution," dramatically altering the field of communication and especially the field of journalism. According to Ramonet (2001), once the domain of communication takes its place at the center of a digital society, it invades the spheres of culture and journalism, now subjected to the "tyranny of communication." This invasion of the practice of journalism explains how the news business is transformed from something sparse and parsimonious into something extremely rapid and marked by excess, and, more importantly even, something that is no longer valuable in and of itself, but a commodity subject to the fluctuations of the market. Thus, "what gives value to a news report is the number of people likely to be interested in it, even as this factor is wholly unrelated to whether it is true" (Ramonet 2001, p56). This transformation has imposed a kind of journalism that favors instantaneous information and transforms a journalist into a "reporter of the instantaneous," a mere conduit that "reports on events immediately as they occur" and who "lacks the time to filter and compare, because taking the time to do so will mean he'll be beat out by his colleagues" (Ramonet 2001, pp56–57).

But when it comes to the critical impact of digitization upon the practice of journalism, we must not limit ourselves to considering the dominance and concentration of power of large and burgeoning business and multimedia conglomerates, or the development of new forms of hyper-textual communication that use innovative and convergent media to reach greater audiences faster and further afield. Nor is it enough to acknowledge the important contribution that digital interactivity offers some[3] members of a digital society: multiple opportunities for expressing in fairly restricted spaces or with a certain amount of advertising what they believe they need to communicate, as well as the choice of accessing information from among the multiple sources afforded by digital networks (but which, in actuality, are reduced as a result of the increasingly concentrated corporate landscape).

The "Third Environment"

Changes brought about by digitalization are apparent in digital journalism, as well as in the way that consumers use networks to keep informed. Information "earthquakes," which occur with a certain degree of frequency in the "Third Environment,"[4] as Javier Echeverría (1999) has called the Internet, clearly affect traditional forms of media, not only because their content is influenced by digital news, and because professional practices of traditional journalists are evidently affected by new ICTs and by digitalization, but also because the discourse of traditional media, similarly to web sources, no longer has the exclusivity it once had. Today it is relatively easy to express opinions and present information in cyberspace without being limited by the capricious filters of a news agency. Thousands of people write blogs, and many more participate more or less actively, posting their opinions in newsgroups, discussion forums, or chats, stimulating in this way the narrative capacities that we all may exploit. Our digital society is beginning to be aware of this, and even traditional news and entertainment media reflect this reality. But this phenomenon is not limited to a mere awakening of the messy narratives of bored or restless citizens. Numerous social projects as they develop have found in the new ICTs an ideal mechanism to channel their internal and external communications, and initiatives of the most varied sorts have been nurtured by and grown as a result of these new communication paradigms.

The counter-informative nature of such projects may recall the origins of many publications in a pre-journalism era, an essential trait that to some degree was inherited by conventional journalism as the industry was beginning to take shape, and which gave it an adversarial power of its own. As the press became institutionalized in its quest to guard against abuses of power, it paradoxically and gradually began losing this very role as a consequence of its close association with those in power. It is no longer possible today to hold that media exist outside of the political arena, given the increasingly close links of media organizations to different powerful interests, particularly economic ones, which prevents them from claiming that they are able to function as watchdogs for their own excesses. More than ever before, today's journalist represents powerful interests, and does not, as is commonly believed, constitute a *fourth estate* with power of its own but instead represents the power of the second estate, as many authors have observed.

The alternative press of today has a much more engaged and difficult role to play than that of pioneer journalists. If the latter set out to be a watchdog of power, the former must first contend with the hegemonic discourse of "monolithic thinking," increasingly limiting and definitive in its approach. However, although both forms of discourse—informative and counter-informative—have been driven by digital technology, the use of the Internet by civil society has most strengthened and given prominence to the latter, and greater and deeper changes in civil society are to be expected as a result.

Journalism experienced extraordinary changes as a result of the emergence of the so-called "knowledge" or "information society," two of the pretentious names used to refer to today's digital world. Despite the decisive influence of new ICTs in an increasing number of areas, many structural imbalances limit truly universal access to autonomous use of digital resources, necessary for information or knowledge to become the vital axes of a 21st-century society. There may be far more reasons to think of such axes instead as misinformation or ignorance.

On the other hand, the use of a utopian term such as "Informed Society" presents a number of obstacles. Only one-sixth of the world's population has access today to this network believed to be the paradigm of what is "universal" and involves constraints that are often overlooked. Dominique Wolton (2000, p91) has written extensively on the subject and warns:

> ... an enthusiastic generation dreams of changing power relations. In good faith, this generation believes the Internet may change the operating rules of capitalism. But if the issues of the division of labor, hierarchies, and questions about who's in charge are not addressed, this enthusiasm will turn to disappointment. The experience of interactive Internet users will not do away with questions of hierarchy, power or concentration.

It would seem that the university's use of the Internet as an educational and cultural medium, which came to represent a second life for the Internet, has had a greater impact upon the corporate world than upon the social fabric. Over time we have seen that not all that glitters is gold on the web, and many of the myths that first surrounded it have begun to be debunked. The Internet is unquestionably an extremely powerful tool to disseminate, enrich, and exploit knowledge. However, since accessibility is precisely the key to its great capacity for communication, the question is whether the Internet will fulfill its optimum potential of capturing and disseminating knowledge in a world where the gap is growing between those who have access to ICTs and those who hardly know they exist. Can this true potential develop if the web does not reach equally all sectors of the population? As Jeremy Rifkin (2000, p25) wrote: "the gap between the haves and have-nots is wide, but that between the connected and the unconnected is even greater." Indeed, we live in a society increasingly determined by access: "to have access" is equivalent to what until recently was meant by "to own."

Ultimately, the Internet age has changed some things directly and many more indirectly. Here we will refer to digital or digitized, and global or globalized society, to avoid overstating such terms in this modest chapter. In any event, in spite of the fact that a relative degree of accessibility of ICTs allows alternative messages to influence public opinion more effectively, and thereby change the course of events and the discourse of mainstream media, the remainder of this chapter will show that a dual form of journalism persists, dividing the center from the periphery. The hegemonic power exerted by the center will continue to determine the forms and content of social communication systems.

However, although many observers are tempted to see similarities with Orwell's Big Brother, this process of globalization is also fortunately marked by critical differences. The media are increasingly clustered in fewer blocks, offering more standardized and less varied information; but the resulting audience segmentation reflects another kind of diversity that somehow compensates for the loss of plurality by the offerings of multiple multimedia channels. Moreover, the Internet and new forms of multilateral communication offer a horizon of possibilities that lead us, perhaps too naively and simplistically, to hope for a way out of the "crisis of intelligibility in which our intellectual and conceptual tools do not allow us to understand the new situation" (Ramonet 2001, p60). Indeed, the impact for journalists of new ICTs goes beyond transforming their work or organizational environment. Instead, the reach of these technologies is evident in the enormous changes experienced in the daily work of journalists, and in the corporate concentration of the news industry, which decidedly shapes the news agenda. Digitization acts in more unpredictable ways by offering audiences the possibility of actively or passively participating in information processes in various ways. At present this participation is limited and only marginal, albeit with effects that, on very rare occasions, have led to a *coup d'état* or the toppling of governments. For many authors, such as Javier Echeverría (1999, p315), the real key to the changes wrought by digitization lies here:

> But the fundamental shift is the relationship between readers and distributors of information. Internet is a horizontal network, multidirectional, decentralized and interactive, and the combination of these four properties is unusual among previous news media and communication organizations.

This exchange in which a plurality of perspectives in mainstream media are replaced by greater audience and time diversity, and decentralized interactivity, however, does not mean that the communication system that is emerging in an age of globalization will favor the social bonds necessary for democracy, especially one that is truly participatory. Dominique Wolton (2000, p36) notes:

> Each individual is connected from his or her personal interests. But how does one shift from the personal to the greater public interest? In fact, communities can live together in perfect mutual indifference.

Considering the established role that media institutions have played as creators of social links, Echeverría contradicts himself when he equates the Internet with traditional news media, despite his insistence on saying that the Internet is not only a medium for information delivery. This is instead new territory, virtual in character, which has only just begun to be colonized, and in which nearly all indications suggest, regrettably, that large-scale virtual landowners are laying claim to most of cyberspace, however infinite this landscape may seem and despite the apparent contradiction this suggests. For Wolton (2000, p66), this distinctive character of the Internet is what holds enormous potential for media; but this potential does little, at least directly, to create the social bonds that have been dependent on media institutions of a broader orientation:

> Internet, as an interactive and automated information system, derives its strength from the fact that it is not a medium of communication: it is messages sent in all directions by anyone, to be picked up by anyone and organized by no one.

In this new territory, however, unlike in the other two environments which have already been colonized, new and very different forces operate, making it, as was already noted, something more than an object for power or control: a hoped-for Arcadia of freedom and equality. Like the suggestive gravitational shift of a distant planet, the interactive, hyper-textual, decentralized, multi-directional, horizontal, and universal virtues of cyberspace all play a decisive role in the development of new social initiatives for emancipation—often frustrated when we forget that we still live on planet Earth. In point of fact, scarcely 1 billion people have full access to this new environment. The remaining 5.5 billion, however, are still reached more or less directly by echoes emanating from cyberspace, so that many such initiatives will also affect the first and second environments rather more than is sometimes believed.

In any case, we must acknowledge the importance of technology in the "media-morphosis" that we are witnessing. This seems to be a much wider and deeper phenomenon than that described by Roger F. Fidler (1998): affecting more than the established media institutions as a result of the appearance of new media. Similarly, we should measure to what extent the alleged new communication paradigm will make it possible to develop further the older model of representative democracy towards greater debate and more enriching methods of participation. Of course, we should refrain from adopting too enthusiastic or even naive a position as Negroponte's (2000), and also avoid falling into apocalyptic attitudes of those who would sign death certificates for journalism or journalists, despite the relative sophistication of some such diagnoses. We must also avoid an excessive idealization of ICTs as tools for social liberation, and not settle either for a relativistic position. Technology should be granted the role that it rightfully deserves in shaping the function and evolution of a social system of communication.

An uncritical perspective of history could see the new technological paradigm as the capitalist system's response to its own internal contradictions, which became evident from the early 1970s onwards, and, at the same, as a defensive reaction to the challenge posed by the Soviet Union's nuclear and space programs. However, despite the fact that ICTs emerge during economic crises:

> ... this synchronicity is too exact, and the "technological adjustment" this calls for is too rapid and too mechanical, when we consider the lessons of the Industrial Revolution and other historical processes of technological change, which show that even when economic, industrial and technological paths are related, they interact imperfectly and shift slowly.
>
> Castells (1998, pp77–78)

Against the new paradigm that Castells identifies, other authors such as Gaëtan Tremblay argue that what underlies this digitalization process is not a revolution. This is because, as we have learned following the demise of the "dotcom bubble," the so-called new economy has maintained, more than ever, the principal rationale of the capitalist system: pursue the greatest possible profit, and the so-called information society is nothing more than an ideological model in which information is still the currency of power. According to Tremblay (2003, p18), "Information Society theory fails to free itself from technological determinism, even when it claims to be distant from it, as Castells argued."

Certainly, a researcher seeking to understand "Internet Society" will have trouble solving the dilemma of the technological origins of a digitized society—the most neutral term we have found—regardless of whether this concept signals the beginning of a new socio-economic system or a sophisticated and late form of capitalism.

Despite the intensity of this debate, the fact remains that, far from imaginary, these are real problems, at least as regards the questions raised by this study. Clearly, technology has become a fundamental feature of 21st-century societies, and the key to understanding many of the changes brought about in culture and communication for a globalized society. More importantly, governments and supranational institutions have much to say and do in terms of technology despite the considerable amount of power that they have ceded to market forces. Civil society has given birth to multitudes of initiatives for social emancipation; their viability and very existence depend on how governments regulate technology. Even though they employ sophisticated technological tools, such initiatives depend on supportive public institutions to be relevant and effective in a cyberspace subject to market criteria.

Conclusion and Analysis Proposals

The intense effervescence of this world of communications leads us to ask questions related to the evolution of the role of conventional journalism from its original activity, reserved exclusively for information professionals acting as their representatives or vicarious agents. As was noted from the beginning of this chapter, the biggest beneficiaries of this globalized movement have been major media companies, who see in this Third Environment of the Internet great opportunities to further reduce costs and increase profits for their investors, at the expense of quality of journalistic production, now much diminished and not fully adapted to the logic of the Internet (Herbert 2000, pp169–184; Bruns 2005, pp310–311; Deuze 2009, pp24–26).

The incorporation of new ICTs within journalists' workplaces has radically affected the pace of their work and, to some extent, the very way in which journalists work. However, although audiences are engaged in many participatory initiatives, and even though the convergence of technology has changed many aspects of journalists' immediate work environment, professional journalists working in mainstream media have barely seen any transformation of the basic tenets of their work, which perhaps more than ever before is essentially a solitary activity not easily relegated to others until publication or broadcast. Although most media companies have assimilated the web environment as the platform for their work, there is no evidence that the production models of conventional media organizations have been transformed by the web beyond the greater speed and concentration of multiple functions the Internet brought to the newsrooms of newspapers, radios, televisions, and news agencies (McChesney 2004, pp57–89; Bruns 2005, p309). These media institutions are now capable of operating at previously unimaginable speeds, drawing on infinitely greater and more accessible resources, and even with some degree of operational coordination as a result of ICTs. However, the essential character of this work remains almost wholly unaffected.

A journalist today is at the crossroads of having had to become a "jack of all trades" but left with little or no capacity to reflect critically, facing an audience very different from just 15 years ago, and almost entirely dependent on information from the dominant discourse of mainstream media. An overdependence on the technical tools of digital technology contributes to maintaining static patterns and practices for professional journalists in an environment whose sophistication, in the end, does not depend on the tools employed, but on the innovative applications of technology that will meet the needs of society. Although such audience demands are increasingly influenced and set by market criteria, society still often finds ways to use technology that the market did not foresee.

But citizens of this form of globalization also, and perhaps more generally, find great benefit to living in an environment that offers all kinds of opportunities and resources for social participation, collective intelligence (de Kerckhove 1999; Deuze 2009), and diverse sources of information, especially those that are distinct from established mainstream media. Such opportunities may, in fact, lead to and shape actual engaged audiences who are able to challenge the work of journalists and media organizations. In this way, the energy for informing and reporting shown by audiences in recent years as a result of new digital possibilities is transforming alternative venues for information and counter-information into required reference points for professional journalists.

On the other hand, this communication also reveals a spiraling "cultural chaos" (McNair 2006) that comes close to what in Spanish is known quite literally as a "cage of crickets" (*jaula de grillos*), where "everybody talks and nobody listens." We live in an age of "alter-mediation" marked by the transformation of the traditional model of social mediation in two opposite and complementary directions: first, established media organizations have lost some of the traditional role they played during the past 300 years as thoughtful intermediaries between society and contemporary events; secondly, the proliferation of new actors in the arena of journalism, regardless of expertise, introduces new "remediation processes" (Bruns 2005, pp311–312), which will significantly change the play's script and also the methods for its interpretation.

Blogs, as well as list-serves, online forums, and alternative information websites, open new processes for remediation that inevitably transform the role of mainstream media in digital society. This transformation has even led to predictions about the end of mass communication, about a transition to "mass self-communication" society, according to Manuel Castells (2009). This results from the dynamic of continually fragmented audiences and increasing options for personalizing content, which presents the risk that this would lead to an undesirable unraveling of the social fabric due to the loss of shared common information reference points (Wolton 2000).

While clearly there has been a noticeable displacement of the role of traditional corporate media, we should not stop thinking of mainstream media as the major shapers of the information agenda. They still decide what is newsworthy and what topics should be of interest to society. Moreover, questions of interactivity, alter-mediation, and personalization involve social forces that could even be the key to saving news professionals from the fate of those who are victimized by digitization, allowing them to rediscover their role in society (Bruns 2005; Deuze 2009). Faced with such a noisy landscape, most journalists continue showing great caution and mistrust of the new net players, and their complex forms of mediation (García de Madariaga 2008, 2010). This does not prevent them from often doing just the opposite and impatiently publishing information without properly framing the information. We are left to conclude, in keeping with the third assumption stated in this chapter, that journalists are failing to sufficiently exploit the Internet's potential for their work not only because of too much pressure from the bottom line and their office culture, but also because they are little prepared to know how to use ICTs, and they do not know what advantages they could bring to their work, especially as regards a greater level of professional autonomy.

However, from the operational point of view, journalists could experience changes in their work if they included the tools for collaboration that digital networks provide, as social movements have done. At present, collaboration in the newsrooms has only occasionally meant coordinating tasks, as was noted earlier. We cannot conclude here either that the journalistic practices that emerged from the use of digital resources led to more updated skills for professional journalism and better equipped it to respond to the needs and demands from a digital society.

Developing a more collaborative form of journalism is consistent with the hypothesis that describes a journalist of the future as a non-specialized professional, with certain specific skills, but whose key competency is to be a mediator of a communication system consisting of an infinite number of non-professionals who are expert sources of information (Bowman and Willis 2003; Gillmor 2004; García de Madariaga 2008), and who is able to act as a conveyor belt promoting the flow of information in much more complex processes than his predecessors.

Nevertheless, perhaps due to the suspicion and fear that journalists have regarding new sources of information on the web and "intruder reporters," many professionals advocate a subject specialization as a way of differentiating themselves from non-professionals and as a guarantee of their own impartiality and credibility. Still, it bears remembering that a skill for communicating is also a form of specialization.

In such an increasingly fragmented and atomized world, it may be more necessary, defensible, and appropriate than ever for journalists to play a pivotal role of bridging worlds that grow more separate from another the more specialized they become. It is also necessary to educate, which means reporting on what, in principle, may not be of interest to audiences, and which may nonetheless be very valuable to know. For this reason, the role of journalists as guarantors of social bonds as advocated by Wolton seems to be the most appropriate. In contrast to the journalist specialized in a particular subject matter and on different target audiences, we may envision a journalist whose specialty is in the sources of information. Therefore, in order to justify its role in a new interactive arena, journalism may follow two complementary paths: it may strengthen its capacity to interpret events at a highly specialized level and/or act as a key link in a much richer and more complex mechanism than the current assembly-line news production process of conventional media. If journalists are to fulfill their role as communicators, they should be able to draw from the broadest view possible of the events being represented.

The benefit of such experiences of collective intelligence in society as well as in online newsrooms is precisely what leads us to a new hypothesis. If it is true that interactivity allows for greater autonomy, it may also facilitate or even require as essential that the production of news and information is done in a way that most clearly takes into account the concerns of society. Just as a professional journalist must individually adapt to an interactive work scenario, traditional media and teams of reporters must also find interactive practices and methods that allow them to differentiate their work product from that of the new web "*producers*" (Bruns 2005, pp313–315). Of course, this model is in opposition to the most esteemed standards of journalism of the industrial age, which nonetheless failed to meet the challenges presented by a desire for greater social emancipation emerging from different ideological constituencies. We may imagine, then, the new profile of a new kind of journalist, able to adapt to these new circumstances in order to satisfy the need for information of a society that aspires to be democratic.

Notes

1 As Professor Enrique de Aguinaga (2002, pp161–162) reports:

> Eight years ago (1994) the Constitutional Committee of the Congress of Deputies called on fourteen distinguished leaders of the professions of Journalism, Business, Education, and Law, to ask them a question that has for many years been the inquiry of many specialized congresses and forums, with scant results: What does it mean to be a journalist? ... In the final session, the Socialist deputy del Pozo concluded the testimony of the fourteen witnesses by stating the following, which amounts to a judgment in its own right: "It is impossible and undesirable to attempt to define what it means to be a journalist."

2 It cost the legendary *The Times* newspaper a one-year closure to implement the new work structure in 1979 as a result of strong resistance by labor unions before the company was sold to Rupert Murdoch's emerging empire, which then fired 6,000 workers from its four newspapers during a single weekend in 1981.

3 Less than 16 percent of the world's population are users of the Internet (http://www.internetworldstats.com/stats.htm).

4 In this chapter we assume such an environment as the best metaphor to represent what the Internet and its converging technologies imply for transformation—projection, at least—of public space, in general, and the communicative sphere, in particular.

References

Bell, D. (1989) *Las contradicciones culturales del capitalismo*. Madrid: Alianza Universidad.

Boczkowski, P. (2002) "The development and use of online newspapers: What research tells us and what we might want to know," in L. Lievrouw and S. Livingstone (eds) *The Handbook of New Media*. London: Sage Publications, 270–286.

Boczkowski, P. J. (2004) *Digitizing the News: Innovation in Online Newspapers*. Cambridge, MA: MIT Press.

Bowman, S. and C. Willis (2003) *We Media: How the Audiences are Shaping the Future of News and Information*. Media Center, American Press Institute; http://www.hypergene.net/wemedia.

Bruns, A. (2005) *Gatewatching: Collaborative Online News Production*. New York, NY: Peter Lang.

Canel, M. J. and T. Sádaba (1999) "La investigación académica sobre las actitudes profesionales de los periodistas: Una descripción del estado de la cuestión," *Comunicación y Sociedad*, XII(2): 9–32.

Canel, M. J. and J. J. Sánchez-Aranda (1999) "La influencia de las actitudes profesionales de *periodista* español en las noticias," *Analisi: Quaderns de Comunicacioi Cultura*, 23.

Castells, M. (1998) *La era de la información: Economía, sociedad y cultura. Volumen I. La Sociedad Red.* Madrid: Alianza Editorial.
Castells, M. (2009) *Comunicación y poder.* Madrid: Alianza Editorial.
Dader, J. (1992) *El periodista en el espacio público.* Barcelona: Bosch.
de Aguinaga, E. (2002) "El periodista en el umbral del siglo XXI," *Estudios Sobre el Mensaje Periodístico*, 8: 157–170.
de Kerckhove, D. (1999) *Inteligencias en conexión: hacia una sociedad de la web.* Barcelona: Gedisa.
Deuze, M. (2009) "Journalism, citizenship and digital culture," in Z. Papacharissi (ed) *Journalism and Citizenship: New Agendas in Communication.* New York, NY: Routledge.
Echeverría, J. (1999) *Los Señores del aire: Telépolis y el Tercer Entorno.* Barcelona: Destino.
Fidler, R. (1998) *Mediamorfosis: Comprender los nuevos medios.* Buenos Aires: Granica.
Fishman, M. (1983) *La fabricación de la noticia.* Buenos Aires: Tres Tiempos.
Gans, H. (1979) *Deciding What's News: A Study of CBS Evening News, NBC Nightly News, Newsweek, and Time.* New York, NY: Vintage Books.
García de Cortázar, M. and M. A. García de León (eds) (2000) *Profesionales del periodismo: Hombres y mujeres en los medios de comunicación.* Madrid: CIS.
García de Madariaga, J. (2004) "El periodista profesional ante la convergencia tecnológica," in G. Matilla and T. Agustín y Lara (eds) *Convergencia multimedia y alfabetización digital.* Madrid: Universidad Complutense de Madrid.
García de Madariaga, J. (2006) "Del periodismo cívico al participativo: Nuevos medios, viejas inquietudes," *Zer, Revista de Estudios de Comunicación*, 21: 203–217.
García de Madariaga, J. (2008) *El periodismo en el siglo XXI: Una profesión en crisis ante la digitalización.* Madrid: Dykinson.
García de Madariaga, J. (2010) "Online journalists in Spain: The influence of ideology in the professional culture," *JRE, Online Journal IAMCR*, http://www.jre-iamcr.org.
Gillmor, D. (2004) *We the Media. Grassroot Journalism by the People for the People.* Sebastopol: O'Reilly Media Inc., http://www.oreilly.com/catalog/wemedia/book/index.csp.
Herbert, J. (2000) *Journalism in the Digital Age: Theory and Practice for Broadcast, Print and Online Media.* Oxford: Focal Press.
Johnstone, J., E. J. Slawski, and W. Bowman (1976) *The News People.* Urbana, IL: University Press.
Lippmann, W. (2009 [1925]) *The Phantom Public.* Piscataway, NJ: Transaction Publishers.
Marí Sáez, V. M. (1999) *Globalización, nuevas tecnologías y comunicación.* Madrid: Ediciones de la Torre.
Martínez-Nicolás, M., J. García de Madariaga, and F. Tucho (2006) *Digital Journalism in the Face of the Transformation Electoral of Political Communication.* IAMCR Conference, Cairo, Egypt.
McChesney, R. W. (2004) *The Problem of Media: U.S. Communication Politics in the 21st Century.* New York, NY: Monthly Review Press.
McNair, B. (2006) *Cultural Chaos: Journalism, News and Power in a Globalised World.* New York, NY: Routledge.
Negroponte, N. (2000) *El mundo digital: Un futuro que ya ha llegado.* Barcelona: Ediciones B.
Orihuela, J. L. (2002) "Nuevos paradigmas de la comunicación," *Chasqui*, 77, March.

Ortega, F. and M. L. Humanes (2000) *Algo más que periodistas: Sociología de una profesión*. Barcelona: Ariel.

Pablos Coello, J. M. de (2001) *El periodismo herido: Estudios que delatan divorcio entre prensa y sociedad. El País como referente*. Madrid: Foca.

Ramonet, I. (2001) (ed) *La tiranía de la Comunicación*. Madrid: Debate.

Randall, D. (1999) *El Periodista Universal*. Madrid: Siglo Veintiuno Editores.

Rifkin, J. (2000) *La era del acceso: La revolución de la nueva economía*. Barcelona: Paidos.

Sandoval Martín, M. T. (2005) "El periodista digital: Precariedad laboral y las nuevas oportunidades," *Telos: Cuadernos de Comunicación, Tecnología y Sociedad*, 63.

Shoemaker, P. and Reese, S. (1996) *Mediating the Message*. Whiteplains, NY: Longman Publishers.

Tremblay, G. (2003) "La Sociedad de la Información y la nueva economía: Promesas, realidades y faltas de un modelo ideológico," *Telos*, 54: 16–23.

Tuchman, G. (1983) *La producción de la noticia: Estudio sobre la construcción de la realidad*. Barcelona: Gustavo Gili.

Weaver, D. (1998) *The Global Journalist: News People around the World*. Cresskill, NJ: Hampton Press.

Wolf, M. (1996) "Los emisores de noticias en la investigación sobre comunicación," *Zer, Revista de Estudios de Comunicación*, 3.

Wolton, D. (2000) *Sobrevivir a Internet: Conversaciones con Olivier Jay*. Barcelona: Gedisa.

Index